Teacher Edition

Eureka Math
Grade 6
Module 4

Special thanks go to the Gordon A. Cain Center and to the Department of Mathematics at Louisiana State University for their support in the development of *Eureka Math*.

For a free *Eureka Math* Teacher
Resource Pack, Parent Tip
Sheets, and more please
visit www.Eureka.tools

Published by the non-profit Great Minds

Printed in the U.S.A.
This book may be purchased from the publisher at eureka-math.org
10 9 8 7 6 5 4 3 2

ISBN 978-1-63255-388-1

Eureka Math: A Story of Ratios Contributors

Michael Allwood, Curriculum Writer
Tiah Alphonso, Program Manager—Curriculum Production
Catriona Anderson, Program Manager—Implementation Support
Beau Bailey, Curriculum Writer
Scott Baldridge, Lead Mathematician and Lead Curriculum Writer
Bonnie Bergstresser, Math Auditor
Gail Burrill, Curriculum Writer
Beth Chance, Statistician
Joanne Choi, Curriculum Writer
Jill Diniz, Program Director
Lori Fanning, Curriculum Writer
Ellen Fort, Math Auditor
Kathy Fritz, Curriculum Writer
Glenn Gebhard, Curriculum Writer
Krysta Gibbs, Curriculum Writer
Winnie Gilbert, Lead Writer / Editor, Grade 8
Pam Goodner, Math Auditor
Debby Grawn, Curriculum Writer
Bonnie Hart, Curriculum Writer
Stefanie Hassan, Lead Writer / Editor, Grade 8
Sherri Hernandez, Math Auditor
Bob Hollister, Math Auditor
Patrick Hopfensperger, Curriculum Writer
Sunil Koswatta, Mathematician, Grade 8
Brian Kotz, Curriculum Writer
Henry Kranendonk, Lead Writer / Editor, Statistics
Connie Laughlin, Math Auditor
Jennifer Loftin, Program Manager—Professional Development
Nell McAnelly, Project Director
Ben McCarty, Mathematician
Stacie McClintock, Document Production Manager
Saki Milton, Curriculum Writer
Pia Mohsen, Curriculum Writer
Jerry Moreno, Statistician
Ann Netter, Lead Writer / Editor, Grades 6–7
Sarah Oyler, Document Coordinator
Roxy Peck, Statistician, Lead Writer / Editor, Statistics
Terrie Poehl, Math Auditor
Kristen Riedel, Math Audit Team Lead
Spencer Roby, Math Auditor
Kathleen Scholand, Math Auditor
Erika Silva, Lead Writer / Editor, Grade 6–7
Robyn Sorenson, Math Auditor
Hester Sutton, Advisor / Reviewer Grades 6–7
Shannon Vinson, Lead Writer / Editor, Statistics
Allison Witcraft, Math Auditor

Julie Wortmann, Lead Writer / Editor, Grade 7
David Wright, Mathematician, Lead Writer / Editor, Grades 6–7

This page intentionally left blank

Table of Contents[1]

Expressions and Equations

---

[1]Each lesson is ONE day, and ONE day is considered a 45-minute period.

Grade 6 • Module 4

Expressions and Equations

OVERVIEW

In Module 4, students extend their arithmetic work to include using letters to represent numbers. Students understand that letters are simply "stand-ins" for numbers and that arithmetic is carried out exactly as it is with numbers. Students explore operations in terms of verbal expressions and determine that arithmetic properties hold true with expressions because nothing has changed—they are still doing arithmetic with numbers. Students determine that letters are used to represent specific but unknown numbers and are used to make statements or identities that are true for all numbers or a range of numbers. Students understand the importance of specifying units when defining letters. Students say "Let K represent Karolyn's weight in pounds" instead of "Let K represent Karolyn's weight" because weight cannot be a specific number until it is associated with a unit, such as pounds, ounces, or grams. They also determine that it is inaccurate to define K as Karolyn because Karolyn is not a number. Students conclude that in word problems, each letter (or variable) represents a number, and its meaning is clearly stated.

To begin this module, students are introduced to important identities that are useful in solving equations and developing proficiency with solving problems algebraically. In Topic A, students understand the relationships of operations and use them to generate equivalent expressions (**6.EE.A.3**). By this time, students have had ample experience with the four operations since they have worked with them from kindergarten through Grade 5 (**1.OA.B.3, 3.OA.B.5**). The topic opens with the opportunity to clarify those relationships, providing students with the knowledge to build and evaluate identities that are important for solving equations. In this topic, students discover and work with the following identities: $w - x + x = w, w + x - x = w,$ $a \div b \cdot b = a, a \cdot b \div b = a$ (when $b \neq 0$), and $3x = x + x + x$. Students also discover that if $12 \div x = 4,$ then $12 - x - x - x - x = 0$.

In Topic B, students experience special notations of operations. They determine that $3x = x + x + x$ is not the same as x^3, which is $x \cdot x \cdot x$. Applying their prior knowledge from Grade 5, where whole number exponents were used to express powers of ten (**5.NBT.A.2**), students examine exponents and carry out the order of operations, including exponents. Students demonstrate the meaning of exponents to write and evaluate numerical expressions with whole number exponents (**6.EE.A.1**).

Students represent letters with numbers and numbers with letters in Topic C. In past grades, students discovered properties of operations through example (**1.OA.B.3, 3.OA.B.5**). Now, they use letters to represent numbers in order to write the properties precisely. Students realize that nothing has changed because the properties still remain statements about numbers. They are not properties of letters; nor are they new rules introduced for the first time. Now, students can extend arithmetic properties from manipulating numbers to manipulating expressions. In particular, they develop the following identities: $a \cdot b = b \cdot a, a + b = b + a, g \cdot 1 = g, g + 0 = g, g \div 1 = g, g \div g = 1,$ and $1 \div g = \frac{1}{g}.$ Students understand that a letter in an expression represents a number. When that number replaces that letter, the expression can be evaluated to one number. Similarly, they understand that a letter in an expression can represent a number. When that number is replaced by a letter, an expression is stated (**6.EE.A.2**).

In Topic D, students become comfortable with new notations of multiplication and division and recognize their equivalence to the familiar notations of the prior grades. The expression $2 \times b$ is exactly the same as $2 \cdot b$, and both are exactly the same as $2b$. Similarly, $6 \div 2$ is exactly the same as $\frac{6}{2}$. These new conventions are practiced to automaticity, both with and without variables. Students extend their knowledge of GCF and the distributive property from Module 2 to expand, factor, and distribute expressions using new notation (**6.NS.B.4**). In particular, students are introduced to factoring and distributing as algebraic identities. These include $a + a = 2 \cdot a = 2a$, $(a + b) + (a + b) = 2 \cdot (a + b) = 2(a + b) = 2a + 2b$, and $a \div b = \frac{a}{b}$.

In Topic E, students express operations in algebraic form. They read and write expressions in which letters stand for and represent numbers (**6.EE.A.2**). They also learn to use the correct terminology for operation symbols when reading expressions. For example, the expression $\frac{3}{2x-4}$ is read as "the quotient of three and the difference of twice a number and four." Similarly, students write algebraic expressions that record operations with numbers and letters that stand for numbers. Students determine that $3a + b$ can represent the story: "Martina tripled her money and added it to her sister's money" (**6.EE.A.2b**).

Students write and evaluate expressions and formulas in Topic F. They use variables to write expressions and evaluate those expressions when given the value of the variable (**6.EE.A.2**). From there, students create formulas by setting expressions equal to another variable. For example, if there are 4 bags containing c colored cubes in each bag with 3 additional cubes, students use this information to express the total number of cubes as $4c + 3$. From this expression, students develop the formula $t = 4c + 3$, where t is the total number of cubes. Once provided with a value for the amount of cubes in each bag ($c = 12$ cubes), students can evaluate the formula for t: $t = 4(12) + 3$, $t = 48 + 3$, $t = 51$. Students continue to evaluate given formulas such as the volume of a cube, $V = s^3$, given the side length, or the volume of a rectangular prism, $V = l \cdot w \cdot h$, given those dimensions (**6.EE.A.2c**).

In Topic G, students are introduced to the fact that equations have a structure similar to some grammatical sentences. Some sentences are true: "George Washington was the first president of the United States," or "$2 + 3 = 5$." Some are clearly false: "Benjamin Franklin was a president of the United States," or "$7 + 3 = 5$." Sentences that are always true or always false are called *closed sentences*. Some sentences need additional information to determine whether they are true or false. The sentence "She is 42 years old" can be true or false, depending on who "she" is. Similarly, the sentence "$x + 3 = 5$" can be true or false, depending on the value of x. Such sentences are called *open sentences*. An equation with one or more variables is an open sentence. The beauty of an open sentence with one variable is that if the variable is replaced with a number, then the new sentence is no longer open: It is either *clearly true* or *clearly false*. For example, for the open sentence $x + 3 = 5$:

> If x is replaced by 7, the new closed sentence, $7 + 3 = 5$, is false because $10 \neq 5$.
> If x is replaced by 2, the new closed sentence, $2 + 3 = 5$, is true because $5 = 5$.

From here, students conclude that solving an equation is the process of determining the number or numbers that, when substituted for the variable, result in a true sentence (**6.EE.B.5**). In the previous example, the solution for $x + 3 = 5$ is obviously 2. The extensive use of bar diagrams in Grades K–5 makes solving equations in Topic G a fun and exciting adventure for students. Students solve many equations twice, once with a bar diagram and once using algebra. They use identities and properties of equality that were introduced earlier in the module to solve one-step, two-step, and multi-step equations. Students solve problems finding the measurements of missing angles represented by letters (**4.MD.C.7**) using what they learned in Grade 4 about the four operations and what they now know about equations.

In Topic H, students use their prior knowledge from Module 1 to construct tables of independent and dependent values in order to analyze equations with two variables from real-life contexts. They represent equations by plotting the values from the table on a coordinate grid (**5.G.A.1**, **5.G.A.2**, **6.RP.A.3a**, **6.RP.A.3b**, **6.EE.C.9**). The module concludes with students referring to true and false number sentences in order to move from solving equations to writing inequalities that represent a constraint or condition in real-life or mathematical problems (**6.EE.B.5**, **6.EE.B.8**). Students understand that inequalities have infinitely many solutions and represent those solutions on number line diagrams.

The 45-day module consists of 34 lessons; 11 days are reserved for administering the Mid- and End-of-Module Assessments, returning assessments, and remediating or providing further applications of the concepts. The Mid-Module Assessment follows Topic E, and the End-of-Module Assessment follows Topic H.

Focus Standards

Apply and extend previous understandings of arithmetic to algebraic expressions.[2]

6.EE.A.1 Write and evaluate numerical expressions involving whole-number exponents.

6.EE.A.2 Write, read, and evaluate expressions in which letters stand for numbers.

a. Write expressions that record operations with numbers and with letters standing for numbers. *For example, express the calculation "Subtract y from 5" as $5 - y$.*

b. Identify parts of an expression using mathematical terms (sum, term, product, factor, quotient, coefficient); view one or more parts of an expression as a single entity. *For example, describe the expression $2(8 + 7)$ as a product of two factors; view $(8 + 7)$ as both a single entity and a sum of two terms.*

c. Evaluate expressions at specific values of their variables. Include expressions that arise from formulas used in real-world problems. Perform arithmetic operations, including those involving whole-number exponents, in the conventional order when there are no parentheses to specify a particular order (Order of Operations). *For example, use the formulas $V = s^3$ and $A = 6s^2$ to find the volume and surface area of a cube with sides of length $s = 1/2$.*

6.EE.A.3 Apply the properties of operations to generate equivalent expressions. *For example, apply the distributive property to the expression $3(2 + x)$ to produce the equivalent expression $6 + 3x$; apply the distributive property to the expression $24x + 18y$ to produce the equivalent expression $6(4x + 3y)$; apply properties of operations to $y + y + y$ to produce the equivalent expression $3y$.*

---

[2]6.EE.A.2c is also taught in Module 4 in the context of geometry.

6.EE.A.4   Identify when two expressions are equivalent (i.e., when the two expressions name the same number regardless of which value is substituted into them). *For example, the expressions* $y + y + y$ *and* $3y$ *are equivalent because they name the same number regardless of which number* y *stands for.*

Reason about and solve one-variable equations and inequalities.[3]

6.EE.B.5   Understand solving an equation or inequality as a process of answering a question: which values from a specified set, if any, make the equation or inequality true? Use substitution to determine whether a given number in a specified set makes an equation or inequality true.

6.EE.B.6   Use variables to represent numbers and write expressions when solving a real-world or mathematical problem; understand that a variable can represent an unknown number, or, depending on the purpose at hand, any number in a specified set.

6.EE.B.7   Solve real-world and mathematical problems by writing and solving equations of the form $x + p = q$ and $px = q$ for cases in which p, q, and x are all nonnegative rational numbers.

6.EE.B.8   Write an inequality of the form $x > c$ or $x < c$ to represent a constraint or condition in a real-world mathematical problem. Recognize that inequalities of the form $x > c$ or $x < c$ have infinitely many solutions; represent solutions of such inequalities on number line diagrams.

Represent and analyze quantitative relationships between dependent and independent variables.

6.EE.C.9   Use variables to represent two quantities in a real-world problem that change in relationship to one another; write an equation to express one quantity, thought of as the dependent variable, in terms of the other quantity, thought of as the independent variable. Analyze the relationship between the dependent and independent variables using graphs and tables, and relate these to the equation. *For example, in a problem involving motion at constant speed, list and graph ordered pairs of distances and times, and write the equation* $d = 65t$ *to represent the relationship between distance and time.*

Foundational Standards

Understand and apply properties of operations and the relationship between addition and subtraction.

1.OA.B.3   Apply properties of operations as strategies to add and subtract.[4] *Examples: If* $8 + 3 = 11$ *is known, then* $3 + 8 = 11$ *is also known. (Commutative property of addition.) To add* $2 + 6 + 4$, *the second two numbers can be added to make a ten, so* $2 + 6 + 4 = 2 + 10 = 12$. *(Associative property of addition.)*

---

[3]Except for 6.EE.B.8, this cluster is also taught in Module 4 in the context of geometry.
[4]Students need not use formal terms for these properties.

Understand properties of multiplication and the relationship between multiplication and division.

3.OA.B.5   Apply properties of operations as strategies to multiply and divide.[5] *Examples: If* $6 \times 4 = 24$ *is known, then* $4 \times 6 = 24$ *is also known. (Commutative property of multiplication.)* $3 \times 5 \times 2$ *can be found by* $3 \times 5 = 15$*, then* $15 \times 2 = 30$*, or by* $5 \times 2 = 10$*, then* $3 \times 10 = 30$*. (Associative property of multiplication.) Knowing that* $8 \times 5 = 40$ *and* $8 \times 2 = 16$*, one can find* 8×7 *as* $8 \times (5 + 2) = (8 \times 5) + (8 \times 2) = 40 + 16 = 56$*. (Distributive property.)*

Gain familiarity with factors and multiples.

4.OA.B.4   Find all factor pairs for a whole number in the range 1–100. Recognize that a whole number is a multiple of each of its factors. Determine whether a given whole number in the range 1–100 is a multiple of a given one-digit number. Determine whether a given whole number in the range 1–100 is prime or composite.

Geometric measurement: understand concepts of angle and measure angles.

4.MD.C.5   Recognize angles as geometric shapes that are formed wherever two rays share a common endpoint, and understand concepts of angle measurement:

    a.   An angle is measured with reference to a circle with its center at the common endpoint of the rays, by considering the fraction of the circular arc between the points where the two rays intersect the circle. An angle that turns through 1/360 of a circle is called a "one-degree angle," and can be used to measure angles.

    b.   An angle that turns through n one-degree angles is said to have an angle measure of n degrees.

4.MD.C.6   Measure angles in whole-number degrees using a protractor. Sketch angles of specified measure.

4.MD.C.7   Recognize angle measure as additive. When an angle is decomposed into non-overlapping parts, the angle measure of the whole is the sum of the angle measures of the parts. Solve addition and subtraction problems to find unknown angles on a diagram in real world and mathematical problems, e.g., by using an equation with a symbol for the unknown angle measure.

Write and interpret numerical expressions.

5.OA.A.2   Write simple expressions that record calculations with numbers, and interpret numerical expressions without evaluating them. *For example, express the calculation "add 8 and 7, then multiply by 2" as* $2 \times (8 + 7)$*. Recognize that* $3 \times (18932 + 921)$ *is three times as large as* $18932 + 921$*, without having to calculate the indicated sum or product.*

---

[5]Students need not use formal terms for these properties.

Analyze patterns and relationships.

5.OA.B.3 Generate two numerical patterns using two given rules. Identify apparent relationships between corresponding terms. Form ordered pairs consisting of corresponding terms from the two patterns, and graph the ordered pairs on a coordinate plane. *For example, given the rule "Add 3" and the starting number 0, and given the rule "Add 6" and the starting number 0, generate terms in the resulting sequences, and observe that the terms in one sequence are twice the corresponding terms in the other sequence. Explain informally why this is so.*

Understand the place value system.

5.NBT.A.2 Explain patterns in the number of zeros of the product when multiplying a number by powers of 10, and explain patterns in the placement of the decimal point when a decimal is multiplied or divided by a power of 10. Use whole-number exponents to denote powers of 10.

Graph points on the coordinate plane to solve real-world and mathematical problems.

5.G.A.1 Use a pair of perpendicular number lines, called axes, to define a coordinate system, with the intersection of the lines (the origin) arranged to coincide with the 0 on each line and a given point in the plane located by using an ordered pair of numbers, called its coordinates. Understand that the first number indicates how far to travel from the origin in the direction of one axis, and the second number indicates how far to travel in the direction of the second axis, with the convention that the names of the two axes and the coordinates correspond (e.g., x-axis and x-coordinate, y-axis and y-coordinate).

5.G.A.2 Represent real world and mathematical problems by graphing points in the first quadrant of the coordinate plane, and interpret coordinate values of points in the context of the situation.

Understand ratio concepts and use ratio reasoning to solve problems.

6.RP.A.3 Use ratio and rate reasoning to solve real-world and mathematical problems, e.g., by reasoning about tables of equivalent ratios, tape diagrams, double number line diagrams, or equations.

    a. Make tables of equivalent ratios relating quantities with whole-number measurements, find missing values in the tables, and plot the pairs of values on the coordinate plane. Use tables to compare ratios.

    b. Solve unit rate problems including those involving unit pricing and constant speed. *For example, if it took 7 hours to mow 4 lawns, then at that rate, how many lawns could be mowed in 35 hours? At what rate were lawns being mowed?*

Compute fluently with multi-digit numbers and find common factors and multiples.

6.NS.B.4 Find the greatest common factor of two whole numbers less than or equal to 100 and the least common multiple of two whole numbers less than or equal to 12. Use the distributive property to express a sum of two whole numbers 1–100 with a common factor as a multiple of a sum of two whole numbers with no common factor. *For example, express $36 + 8$ as $4(9 + 2)$.*

Module 4:   Expressions and Equations

Focus Standards for Mathematical Practice

MP.2   **Reason abstractly and quantitatively.** Students connect symbols to their numerical referents. They understand exponential notation as repeated multiplication of the base number. Students realize that 3^2 is represented as 3×3, with a product of 9, and explain how 3^2 differs from 3×2, where the product is 6. Students determine the meaning of a variable within a real-life context. They write equations and inequalities to represent mathematical situations. Students manipulate equations using the properties so that the meaning of the symbols and variables can be more closely related to the real-world context. For example, given the expression $12x$ represents how many beads are available to make necklaces, students rewrite $12x$ as $4x + 4x + 4x$ when trying to show the portion each person gets if there are three people or rewrite $12x$ as $6x + 6x$ if there are two people sharing. Students recognize that these expressions are equivalent. Students can also use equivalent expressions to express the area of rectangles and to calculate the dimensions of a rectangle when the area is given. Also, students make connections between a table of ordered pairs of numbers and the graph of those data.

MP.6   **Attend to precision.** Students are precise in defining variables. They understand that a variable represents one number. For example, students understand that in the equation $a + 4 = 12$, the variable a can only represent one number to make the equation true. That number is 8, so $a = 8$. When variables are represented in a real-world problem, students precisely define the variables. In the equation $2w = 18$, students define the variable as weight in pounds (or some other unit) rather than just weight. Students are precise in using operation symbols and can connect between previously learned symbols and new symbols (3×2 can be represented with parentheses, $3(2)$, or with the multiplication dot $3 \cdot 2$; similarly, $3 \div 2$ is also represented with the fraction bar $\frac{3}{2}$). In addition, students use appropriate vocabulary and terminology when communicating about expressions, equations, and inequalities. For example, students write expressions, equations, and inequalities from verbal or written descriptions. "A number increased by 7 is equal to 11" can be written as $x + 7 = 11$. Students refer to $7y$ as a *term* or an *expression,* whereas $7y = 56$ is referred to as an *equation.*

MP.7   **Look for and make use of structure.** Students look for structure in expressions by deconstructing them into a sequence of operations. They make use of structure to interpret an expression's meaning in terms of the quantities represented by the variables. In addition, students make use of structure by creating equivalent expressions using properties. For example, students write $6x$ as $x + x + x + x + x + x$, $4x + 2x$, $3(2x)$, or other equivalent expressions. Students also make sense of algebraic solutions when solving an equation for the value of the variable through connections to bar diagrams and properties. For example, when there are two copies of $a + b$, this can be expressed as either $(a + b) + (a + b)$, $2a + 2b$, or $2(a + b)$. Students use tables and graphs to compare different expressions or equations to make decisions in real-world scenarios. These models also create structure as students gain knowledge in writing expressions and equations.

MP.8 **Look for and express regularity in repeated reasoning.** Students look for regularity in a repeated calculation and express it with a general formula. Students work with variable expressions while focusing more on the patterns that develop than the actual numbers that the variable represents. For example, students move from an expression such as $3 + 3 + 3 + 3 = 4 \cdot 3$ to the general form $m + m + m + m = 4 \cdot m$, or $4m$. Similarly, students move from expressions such as $5 \cdot 5 \cdot 5 \cdot 5 = 5^4$ to the general form $m \cdot m \cdot m \cdot m = m^4$. These are especially important when moving from the general form back to a specific value for the variable.

Terminology

New or Recently Introduced Terms

- **Equation** (An *equation* is a statement of equality between two expressions.)
- **Equivalent Expressions** (Two expressions are *equivalent* if both expressions evaluate to the same number for every substitution of numbers into all the variables in both expressions.)
- **Exponential Notation for Whole Number Exponents** (Let m be a nonzero whole number. For any number a, the expression a^m is the product of m factors of a (i.e., $a^m = \underbrace{a \cdot a \cdots \cdot a}_{m \text{ times}}$). The number a is called the *base,* and m is called the *exponent* or *power* of a.)
- **Expression** (An *expression* is a numerical expression, or it is the result of replacing some (or all) of the numbers in a numerical expression with variables.)
- **Linear Expression** (A *linear expression* is an expression that is equivalent to the sum/difference of one or more expressions where each expression is either a number, a variable, or a product of a number and a variable.)
- **Number Sentence** (A *number sentence* is a statement of equality between two numerical expressions.)
- **Numerical Expression** (A *numerical expression* is a number, or it is any combination of sums, differences, products, or divisions of numbers that evaluates to a number.)
- **Solution of an Equation** (A *solution* to an equation with one variable is a number such that the number sentence resulting from substituting the number for all instances of the variable in both expressions is a true number sentence.

  If an equation has more than one variable, then a solution is an ordered tuple of numbers such that the number sentence resulting from substituting each number from the tuple into all instances of its corresponding variable is a true number sentence.)
- **Truth Values of a Number Sentence** (A number sentence is said to be *true* if both numerical expressions evaluate to the same number; it is said to be *false* otherwise. True and false are called *truth values*.)
- **Value of a Numerical Expression** (The *value of a numerical expression* is the number found by evaluating the expression.)
- **Variable** (A *variable* is a symbol (such as a letter) that is a placeholder for a number.)

Familiar Terms and Symbols[6]

- Distribute
- Expand
- Factor
- Number Sentence
- Product
- Properties of Operations (distributive, commutative, associative)
- Quotient
- Sum
- Term
- True or False Number Sentence
- Variable or Unknown Number

Suggested Tools and Representations

- Bar model
- Geometric figures
- Protractors

Rapid White Board Exchanges

Implementing an RWBE requires that each student be provided with a personal white board, a white board marker, and a means of erasing his or her work. An economic choice for these materials is to place sheets of card stock inside sheet protectors to use as the personal white boards and to cut sheets of felt into small squares to use as erasers.

An RWBE consists of a sequence of 10 to 20 problems on a specific topic or skill that starts out with a relatively simple problem and progressively gets more difficult. The teacher should prepare the problems in a way that allows him or her to reveal them to the class one at a time. A flip chart or PowerPoint presentation can be used, or the teacher can write the problems on the board and either cover some with paper or simply write only one problem on the board at a time.

The teacher reveals, and possibly reads aloud, the first problem in the list and announces, "Go." Students work the problem on their personal white boards as quickly as possible and hold their work up for their teacher to see their answers as soon as they have the answer ready. The teacher gives immediate feedback to each student, pointing and/or making eye contact with the student and responding with an affirmation for correct work, such as "Good job!", "Yes!", or "Correct!", or responding with guidance for incorrect work such as "Look again," "Try again," "Check your work," and so on. In the case of the RWBE, it is not recommended that the feedback include the name of the student receiving the feedback.

---

[6]These are terms and symbols students have seen previously.

If many students have struggled to get the answer correct, go through the solution of that problem as a class before moving on to the next problem in the sequence. Fluency in the skill has been established when the class is able to go through each problem in quick succession without pausing to go through the solution of each problem individually. If only one or two students have not been able to successfully complete a problem, it is appropriate to move the class forward to the next problem without further delay; in this case, find a time to provide remediation to those students before the next fluency exercise on this skill is given.

Assessment Summary

Assessment Type	Administered	Format	Standards Addressed
Mid-Module Assessment Task	After Topic E	Constructed response with rubric	6.EE.A.1, 6.EE.A.2, 6.EE.A.3, 6.EE.A.4
End-of-Module Assessment Task	After Topic H	Constructed response with rubric	6.EE.A.2, 6.EE.B.5, 6.EE.B.6, 6.EE.B.7, 6.EE.B.8, 6.EE.C.9

Mathematics Curriculum

6

GRADE

Topic A

Relationships of the Operations

6.EE.A.3

Focus Standard:	6.EE.A.3	Apply the properties of operations to generate equivalent expressions. *For example, apply the distributive property to the expression $3(2 + x)$ to produce the equivalent expression $6 + 3x$; apply the distributive property to the expression $24x + 18y$ to produce the equivalent expression $6(4x + 3y)$; apply properties of operations to $y + y + y$ to produce the equivalent expression $3y$.*
Instructional Days:	4	
Lesson 1:	The Relationship of Addition and Subtraction (S)[1]	
Lesson 2:	The Relationship of Multiplication and Division (E)	
Lesson 3:	The Relationship of Multiplication and Addition (S)	
Lesson 4:	The Relationship of Division and Subtraction (S)	

Prior to this module, students have worked with numbers and operations from Kindergarten through Grade 5. In Topic A, students further discover and clarify the relationships of the operations using models. From these models, students build and evaluate identities that are useful in solving equations and developing proficiency with solving problems algebraically.

To begin, students use models to discover the relationship between addition and subtraction. In Lesson 1, for example, a model could represent the number three. Students notice that if two are taken away, there is a remainder of one. However, when students replace the two units, they notice the answer is back to the original three. Hence, students first discover the identity $w - x + x = w$ and later discover that $w + x - x = w$.

In Lesson 2, students model the relationship between multiplication and division. They note that when they divide eight units into two equal groups, they find a quotient of four. They discover that if they multiply that quotient by the number of groups, then they return to their original number, eight, and ultimately build the identities $a \div b \cdot b = a$ and $a \cdot b \div b = a$, when $b \neq a$.

---

[1]Lesson Structure Key: **P**-Problem Set Lesson, **M**-Modeling Cycle Lesson, **E**-Exploration Lesson, **S**-Socratic Lesson

Students continue to discover identities in Lesson 3, where they determine the relationship between multiplication and addition. Using tape diagrams from previous modules, students are assigned a diagram with three equal parts, where one part is assigned a value of four. They note that since there are three equal parts, they can add four three times to determine the total amount. They relate to multiplication and note that three groups with four items in each group produces a product of twelve, determining that $3 \cdot g = g + g + g$.

Finally, in Lesson 4, students relate division to subtraction. They notice that dividing eight by two produces a quotient of four. They experiment and find that if they subtract the divisor from the dividend four times (the quotient), they find a remainder of zero. They continue to investigate with other examples and prove that if they continually subtract the divisor from the dividend, they determine a difference, or remainder, of zero. Hence, $12 \div x = 4$ means $12 - x - x - x - x = 0$.

EUREKA
MATH™

Lesson 1: The Relationship of Addition and Subtraction

Student Outcomes

- Students build and clarify the relationship of addition and subtraction by evaluating identities such as $w - x + x = w$ and $w + x - x = w$.

Lesson Notes

Teachers need to create square pieces of paper in order for students to build tape diagrams. Each pair of students needs 10 squares to complete the activities. If the teacher has square tiles, these can be used in place of paper squares.

The template for the squares and other shapes used in the lesson are provided at the end of the lesson. Teachers need to cut out the shapes.

Classwork

Fluency Exercise (5 minutes): Multiplication of Decimals

RWBE: Refer to the Rapid White Board Exchanges sections in the Module Overview for directions to administer an RWBE.

Opening Exercise (5 minutes)

Opening Exercise

    a.   Draw a tape diagram to represent the following expression: $5 + 4$.

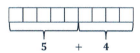

    b.   Write an expression for each tape diagram.

        i.

        ii.

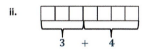

Discuss the answers with the class. If students struggled with the Opening Exercise, provide more examples before moving into the Discussion.

Discussion (15 minutes)

Provide each pair of students with a collection of 10 squares, so they can use these squares to create tape diagrams throughout the lesson.

- If each of the squares represents 1 unit, represent the number 3 using the squares provided.
  -

- Add two more squares to your tape diagram.
  - ▢▢▢▢▢

- Write an expression to represent how we created a tape diagram with five squares.
  - ▢▢▢▢▢

    3 + 2

- Remove two squares from the tape diagram.
  - ▢▢▢

- Alter our original expression $3 + 2$ to create an expression that represents what we did with the tape diagram.
  - $3 + 2 - 2$

- Evaluate the expression.
  - 3

- Let's start a new diagram. This time, create a tape diagram with six squares.
  - ▢▢▢▢▢▢

- Use your squares to demonstrate the expression $6 + 4$.
  - ▢▢▢▢▢▢▢▢▢▢

- Remove four squares from the tape diagram.
  - ▢▢▢▢▢▢

- Alter the expression $6 + 4$ to create an expression to represent the tape diagram.
  - $6 + 4 - 4$

- How many squares are left on your desk?
  - 6

- Evaluate the expression.
  - 6

- How many squares did we start with?
  - 6

- What effect did adding four squares and then subtracting the four squares have on the number of squares?
  - *Adding and then subtracting the same number of squares resulted in the same number that we started with.*

- What if I asked you to add 215 squares to the six squares we started with and then subtract 215 squares? Do you need to actually add and remove these squares to know what the result will be? Why is that?
  - *We do not actually need to do the addition and subtraction because we now know that it will result in the same amount of squares that we started with; when you add and then subtract the same number, the results will be the original number.*

EUREKA
MATH™

- What do you notice about the expressions we created with the tape diagrams?
  - *Possible answer: When we add one number and then subtract the same number, we get our original number.*
- Write an equation, using variables, to represent what we just demonstrated with tape diagrams. Remember that a variable is a letter that represents a number. Use the shapes provided to create tape diagrams to demonstrate this equation.

Provide students time to work with their partners to write an equation.

MP.2

  - *Possible answer: $w + x - x = w$. Emphasize that both w's represent the same number, and the same rule applies to the x's.*
- Why is the equation $w + x - x = w$ called an *identity*?
  - *Possible answer: The equation is called an identity because the variables can be replaced with any numbers, and after completing the operations, I returned to the original value.*

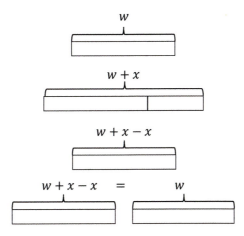

Exercises (12 minutes)

Students use their knowledge gained in the Discussion to create another equation using identities. Allow students to continue to work with their partners and 10 squares.

> *Scaffolding:*
> The exercise could be completed as a class if students are struggling with the concept.

MP.7

Exercises

1. Predict what will happen when a tape diagram has a large number of squares, some squares are removed, and then the same amount of squares are added back on.

   Possible answer: When some squares are removed from a tape diagram, and then the same amount of squares are added back on, the tape diagram will end up with the same amount of squares that it started with.

2. Build a tape diagram with 10 squares.

   a. Remove six squares. Write an expression to represent the tape diagram.

      $10 - 6$

   b. Add six squares onto the tape diagram. Alter the original expression to represent the current tape diagram.

      $10 - 6 + 6$

c.   Evaluate the expression.

10

3.   Write an equation, using variables, to represent the identities we demonstrated with tape diagrams.

Possible answer: $w - x + x = w$

4.   Using your knowledge of identities, fill in each of the blanks.

a. $4 + 5 - \underline{\quad} = 4$

5

b. $25 - \underline{\quad} + 10 = 25$

10

c. $\underline{\quad} + 16 - 16 = 45$

45

d. $56 - 20 + 20 = \underline{\quad}$

56

5.   Using your knowledge of identities, fill in each of the blanks.

a. $a + b - \underline{\quad} = a$

b

b. $c - d + d = \underline{\quad}$

c

c. $e + \underline{\quad} - f = e$

f

d. $\underline{\quad} - h + h = g$

g

EUREKA MATH

Closing (3 minutes)

- In every problem we did today, why did the final value of the expression equal the initial expression?
  - *The overall change to the expression was* 0.
- Initially, we added an amount and then subtracted the same amount. Later in the lesson, we subtracted an amount and then added the same amount. Did this alter the outcome?
  - *This did not alter the outcome; in both cases, we still ended with our initial value.*
- Why were we able to evaluate the final expression even when we did not know the amount we were adding and subtracting?
  - *If we add and subtract the same value, it is similar to adding* 0 *to an expression because the two numbers are opposites, which have a sum of* 0.

Exit Ticket (5 minutes)

©2015 Great Minds. eureka-math.org
G6-M4-TE-B4-1.3.1-01.2016

Name _____   Date _____

Lesson 1: The Relationship of Addition and Subtraction

Exit Ticket

1. Draw tape diagrams to represent each of the following number sentences.

    a. $3 + 5 - 5 = 3$

    b. $8 - 2 + 2 = 8$

2. Fill in each blank.

    a. $65 + \underline{\quad\quad} - 15 = 65$

    b. $\underline{\quad\quad} + g - g = k$

    c. $a + b - \underline{\quad\quad} = a$

    d. $367 - 93 + 93 = \underline{\quad\quad}$

EUREKA
MATH™

Exit Ticket Sample Solutions

1. Draw a series of tape diagrams to represent the following number sentences.

   a. $3 + 5 - 5 = 3$

   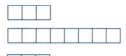

   b. $8 - 2 + 2 = 8$

2. Fill in each blank.

   a. $65 + \underline{\quad} - 15 = 65$

   15

   b. $\underline{\quad} + g - g = k$

   k

   c. $a + b - \underline{\quad} = a$

   b

   d. $367 - 93 + 93 = \underline{\quad}$

   367

Problem Set Sample Solutions

1. Fill in each blank.

   a. $\underline{\quad} + 15 - 15 = 21$ b. $450 - 230 + 230 = \underline{\quad}$ c. $1289 - \underline{\quad} + 856 = 1289$

   21   **450**   **856**

2. Why are the equations $w - x + x = w$ and $w + x - x = w$ called *identities*?

   Possible answer: These equations are called identities because the variables can be replaced with any numbers, and after completing the operations, I returned to the original value.

Multiplication of Decimals

Progression of Exercises

1. $0.5 \times 0.5 =$

   0.25

2. $0.6 \times 0.6 =$

   0.36

3. $0.7 \times 0.7 =$

   0.49

4. $0.5 \times 0.6 =$

   0.3

5. $1.5 \times 1.5 =$

   2.25

6. $2.5 \times 2.5 =$

   6.25

7. $0.25 \times 0.25 =$

   0.0625

8. $0.1 \times 0.1 =$

   0.01

9. $0.1 \times 123.4 =$

   12.34

10. $0.01 \times 123.4 =$

    1.234

EUREKA
MATH™

$$w + x$$

$$w$$ $$x$$

$$w$$ $$x$$

$$w + x$$

$$w$$ $$x$$

$$w$$ $$x$$

Lesson 1:   The Relationship of Addition and Subtraction

EUREKA
MATH™

Lesson 2: The Relationship of Multiplication and Division

Student Outcomes

- Students build and clarify the relationship of multiplication and division by evaluating identities such as $a \div b \cdot b = a$ and $a \cdot b \div b = a$.

Lesson Notes

Students use the squares that were used in Lesson 1; however, each pair of students should receive 20 squares for this lesson. Also, students need large paper to complete the Exploratory Challenge.

Classwork

Fluency Exercise (5 minutes): Division of Fractions I

Sprint: Refer to the Sprints and Sprint Delivery Script sections in the Module Overview for directions to administer a Sprint.

Opening (2 minutes)

MP.2 Remind students of the identities they learned the previous day. Discuss the relationship between addition and subtraction. Inform students that the relationship between multiplication and division is discussed today. Have students make predictions about this relationship using their knowledge gained in the previous lesson.

Opening Exercise (5 minutes)

Opening Exercise

Draw a pictorial representation of the division and multiplication problems using a tape diagram.

a. $8 \div 2$

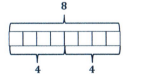 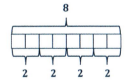

b. 3×2

Discussion (optional—see Scaffolding notes)

Provide each pair of students with a collection of 20 squares, which they use to create tape diagrams throughout the lesson.

> **Scaffolding:**
>
> The Discussion is provided if students struggled during Lesson 1. If the Discussion is included in the lesson, the Exploratory Challenge is shortened because students only develop one number sentence.

- Build a tape diagram to represent 9 units.
    -

- Divide the 9 units into three equal groups.
    -

- Write an expression to represent the process you modeled with the tape diagram.
    - $9 \div 3$

- Evaluate the expression.
    - 3

- Use your squares to demonstrate what it would look like to multiply 3 by 3.
    -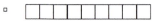

- Alter our original expression, $9 \div 3$, to create an expression that represents what we did with the tape diagram.
    - $9 \div 3 \times 3$

- Evaluate the expression.
    - 9

- What do you notice about the expression of the tape diagram?
    - *Possible answer: When we divide by one number and then multiply by the same number, we end up with our original number.*

 MP.7

- Write an equation, using variables, to represent the identities we demonstrated with tape diagrams. Draw a series of tape diagrams to demonstrate this equation.
    - *Provide students time to work in pairs.*
    - *Possible answer: $a \div b \times b = a$. Emphasize that both a's represent the same number, and the same rule applies to the b's.*

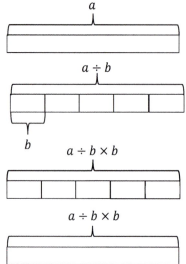

©2015 Great Minds. eureka-math.org
G6-M4-TE-B4-1.3.1-01.2016

Exploratory Challenge (23 minutes)

Students work in pairs or small groups to determine equations to show the relationship between multiplication and division. They use tape diagrams to provide support for their findings.

Scaffolding:

If students struggle with getting started, show them the identity equations for addition and subtraction learned in Lesson 1.

Exploratory Challenge

Work in pairs or small groups to determine equations to show the relationship between multiplication and division. Use tape diagrams to provide support for your findings.

1. Create two equations to show the relationship between multiplication and division. These equations should be identities and include variables. Use the squares to develop these equations.

2. Write your equations on large paper. Show a series of tape diagrams to defend each of your equations.

    Only one number sentence is shown there; the second number sentence and series of tape diagrams are included in the optional Discussion.

MP.7

Possible answer: $a \times b \div b = a$

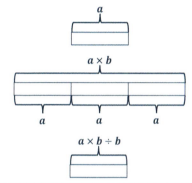

Possible answer: $a \div b \times b = a$

When students complete their work on the large paper, hang the papers around the room. Provide time for students to walk around and critique their peers' work. While examining the other posters, students should be comparing the equations and tape diagrams to their own.

MP.3

Use the following rubric to critique other posters.

1. Name of the group you are critiquing

2. Equation you are critiquing

3. Whether or not you believe the equations are true and reasons why

---

Closing (5 minutes)

- What did you determine about the relationship of multiplication and division?
    - *When a number is multiplied and divided by the same number, the result is the original number.*
- What equations can be used to show the relationship of multiplication and division?
    - $a \div b \cdot b = a$ *and* $a \cdot b \div b = a$

Exit Ticket (5 minutes)

Name _____   Date _____

Lesson 2: The Relationship of Multiplication and Division

1. Fill in the blanks to make each equation true.

    a. $12 \div 3 \times$ _____ $= 12$

    b. $f \times h \div h =$ _____

    c. $45 \times$ _____ $\div 15 = 45$

    d.   _____ $\div r \times r = p$

2. Draw a series of tape diagrams to represent the following number sentences.

    a. $12 \div 3 \times 3 = 12$

    b. $4 \times 5 \div 5 = 4$

©2015 Great Minds. eureka-math.org
G6-M4-TE-B4-1.3.1-01.2016

EUREKA MATH™

Exit Ticket Sample Solutions

1. Fill in the blanks to make each equation true.

    a. $12 \div 3 \times$ ___ $= 12$

        3

    b. $f \times h \div h =$ ___

        f

    c. $45 \times$ ___ $\div 15 = 45$

        15

    d. ___ $\div r \times r = p$

        p

2. Draw a series of tape diagrams to represent the following number sentences.

    a. $12 \div 3 \times 3 = 12$

    b. $4 \times 5 \div 5 = 4$

Problem Set Sample Solutions

1. Fill in each blank to make each equation true.

    a. $132 \div 3 \times 3 =$ ___   b. ___ $\div 25 \times 25 = 225$

        132   225

    c. $56 \times$ ___ $\div 8 = 56$ d. $452 \times 12 \div$ ___ $= 452$

        8   12

2. How is the relationship of addition and subtraction similar to the relationship of multiplication and division?

    Possible answer: Both relationships create identities.

Number Correct: _____

Division of Fractions—Round 1

Directions: Evaluate each expression and simplify. Use blank spaces to create like units, where applicable.

1.	9 ones ÷ 3 ones		23.	$\dfrac{6}{10} \div \dfrac{4}{10}$	
2.	$9 \div 3$		24.	$\dfrac{6}{10} \div \dfrac{2}{5} = \dfrac{6}{10} \div \dfrac{}{10}$	
3.	9 tens ÷ 3 tens		25.	$\dfrac{10}{12} \div \dfrac{5}{12}$	
4.	$90 \div 30$		26.	$\dfrac{5}{6} \div \dfrac{5}{12} = \dfrac{}{12} \div \dfrac{5}{12}$	
5.	9 hundreds ÷ 3 hundreds		27.	$\dfrac{10}{12} \div \dfrac{3}{12}$	
6.	$900 \div 300$		28.	$\dfrac{10}{12} \div \dfrac{1}{4} = \dfrac{10}{12} \div \dfrac{}{12}$	
7.	9 halves ÷ 3 halves		29.	$\dfrac{5}{6} \div \dfrac{3}{12} = \dfrac{}{12} \div \dfrac{3}{12}$	
8.	$\dfrac{9}{2} \div \dfrac{3}{2}$		30.	$\dfrac{5}{10} \div \dfrac{2}{10}$	
9.	9 fourths ÷ 3 fourths		31.	$\dfrac{5}{10} \div \dfrac{1}{5} = \dfrac{5}{10} \div \dfrac{}{10}$	
10.	$\dfrac{9}{4} \div \dfrac{3}{4}$		32.	$\dfrac{1}{2} \div \dfrac{2}{10} = \dfrac{}{10} \div \dfrac{2}{10}$	
11.	$\dfrac{9}{8} \div \dfrac{3}{8}$		33.	$\dfrac{1}{2} \div \dfrac{2}{4}$	
12.	$\dfrac{2}{3} \div \dfrac{1}{3}$		34.	$\dfrac{3}{4} \div \dfrac{2}{8}$	
13.	$\dfrac{1}{3} \div \dfrac{2}{3}$		35.	$\dfrac{1}{2} \div \dfrac{3}{8}$	
14.	$\dfrac{6}{7} \div \dfrac{2}{7}$		36.	$\dfrac{1}{2} \div \dfrac{1}{5} = \dfrac{}{10} \div \dfrac{}{10}$	
15.	$\dfrac{5}{7} \div \dfrac{2}{7}$		37.	$\dfrac{2}{4} \div \dfrac{1}{3}$	
16.	$\dfrac{3}{7} \div \dfrac{4}{7}$		38.	$\dfrac{1}{4} \div \dfrac{4}{6}$	
17.	$\dfrac{6}{10} \div \dfrac{2}{10}$		39.	$\dfrac{3}{4} \div \dfrac{2}{6}$	
18.	$\dfrac{6}{10} \div \dfrac{4}{10}$		40.	$\dfrac{5}{6} \div \dfrac{1}{4}$	
19.	$\dfrac{6}{10} \div \dfrac{8}{10}$		41.	$\dfrac{2}{9} \div \dfrac{5}{6}$	
20.	$\dfrac{7}{12} \div \dfrac{2}{12}$		42.	$\dfrac{5}{9} \div \dfrac{1}{6}$	
21.	$\dfrac{6}{12} \div \dfrac{9}{12}$		43.	$\dfrac{1}{2} \div \dfrac{1}{7}$	
22.	$\dfrac{4}{12} \div \dfrac{11}{12}$		44.	$\dfrac{5}{7} \div \dfrac{1}{2}$	

EUREKA
MATH™

Division of Fractions—Round 1 [KEY]

Directions: Evaluate each expression and simplify. Use blank spaces to create like units, where applicable.

1.	9 ones ÷ 3 ones	$\frac{9}{3}=3$	23.	$\frac{6}{10} \div \frac{4}{10}$	$\frac{6}{4}=1\frac{1}{2}$
2.	$9 \div 3$	$\frac{9}{3}=3$	24.	$\frac{6}{10} \div \frac{2}{5} = \frac{6}{10} \div \frac{\ }{10}$	$\frac{6}{4}=1\frac{1}{2}$
3.	9 tens ÷ 3 tens	$\frac{9}{3}=3$	25.	$\frac{10}{12} \div \frac{5}{12}$	$\frac{10}{5}=2$
4.	$90 \div 30$	$\frac{9}{3}=3$	26.	$\frac{5}{6} \div \frac{5}{12} = \frac{\ }{12} \div \frac{5}{12}$	$\frac{10}{5}=2$
5.	9 hundreds ÷ 3 hundreds	$\frac{9}{3}=3$	27.	$\frac{10}{12} \div \frac{3}{12}$	$\frac{10}{3}=3\frac{1}{3}$
6.	$900 \div 300$	$\frac{9}{3}=3$	28.	$\frac{10}{12} \div \frac{1}{4} = \frac{10}{12} \div \frac{\ }{12}$	$\frac{10}{3}=3\frac{1}{3}$
7.	9 halves ÷ 3 halves	$\frac{9}{3}=3$	29.	$\frac{5}{6} \div \frac{3}{12} = \frac{\ }{12} \div \frac{3}{12}$	$\frac{10}{3}=3\frac{1}{3}$
8.	$\frac{9}{2} \div \frac{3}{2}$	$\frac{9}{3}=3$	30.	$\frac{5}{10} \div \frac{2}{10}$	$\frac{5}{2}=2\frac{1}{2}$
9.	9 fourths ÷ 3 fourths	$\frac{9}{3}=3$	31.	$\frac{5}{10} \div \frac{1}{5} = \frac{5}{10} \div \frac{\ }{10}$	$\frac{5}{2}=2\frac{1}{2}$
10.	$\frac{9}{4} \div \frac{3}{4}$	$\frac{9}{3}=3$	32.	$\frac{1}{2} \div \frac{2}{10} = \frac{\ }{10} \div \frac{2}{10}$	$\frac{5}{2}=2\frac{1}{2}$
11.	$\frac{9}{8} \div \frac{3}{8}$	$\frac{9}{3}=3$	33.	$\frac{1}{2} \div \frac{2}{4}$	$\frac{2}{2}=1$
12.	$\frac{2}{3} \div \frac{1}{3}$	$\frac{2}{1}=2$	34.	$\frac{3}{4} \div \frac{2}{8}$	3
13.	$\frac{1}{3} \div \frac{2}{3}$	$\frac{1}{2}$	35.	$\frac{1}{2} \div \frac{3}{8}$	$\frac{4}{3}=1\frac{1}{3}$
14.	$\frac{6}{7} \div \frac{2}{7}$	$\frac{6}{2}=3$	36.	$\frac{1}{2} \div \frac{1}{5} = \frac{\ }{10} \div \frac{\ }{10}$	$\frac{5}{2}=2\frac{1}{2}$
15.	$\frac{5}{7} \div \frac{2}{7}$	$\frac{5}{2}=2\frac{1}{2}$	37.	$\frac{2}{4} \div \frac{1}{3}$	$\frac{6}{4}=1\frac{1}{2}$
16.	$\frac{3}{7} \div \frac{4}{7}$	$\frac{3}{4}$	38.	$\frac{1}{4} \div \frac{4}{6}$	$\frac{3}{8}$
17.	$\frac{6}{10} \div \frac{2}{10}$	$\frac{6}{2}=3$	39.	$\frac{3}{4} \div \frac{2}{6}$	$\frac{9}{4}=2\frac{1}{4}$
18.	$\frac{6}{10} \div \frac{4}{10}$	$\frac{6}{4}=1\frac{1}{2}$	40.	$\frac{5}{6} \div \frac{1}{4}$	$\frac{10}{3}=3\frac{1}{3}$
19.	$\frac{6}{10} \div \frac{8}{10}$	$\frac{6}{8}=\frac{3}{4}$	41.	$\frac{2}{9} \div \frac{5}{6}$	$\frac{4}{15}$
20.	$\frac{7}{12} \div \frac{2}{12}$	$\frac{7}{2}=3\frac{1}{2}$	42.	$\frac{5}{9} \div \frac{1}{6}$	$\frac{15}{3}=5$
21.	$\frac{6}{12} \div \frac{9}{12}$	$\frac{6}{9}=\frac{2}{3}$	43.	$\frac{1}{2} \div \frac{1}{7}$	$\frac{7}{2}=3\frac{1}{2}$
22.	$\frac{4}{12} \div \frac{11}{12}$	$\frac{4}{11}$	44.	$\frac{5}{7} \div \frac{1}{2}$	$\frac{10}{7}=1\frac{3}{7}$

Number Correct: _____
Improvement: _____

Division of Fractions—Round 2

Directions: Evaluate each expression and simplify. Use blank spaces to create like units, where applicable.

1.	12 ones ÷ 2 ones	
2.	$12 \div 2$	
3.	12 tens ÷ 2 tens	
4.	$120 \div 20$	
5.	12 hundreds ÷ 2 hundreds	
6.	$1,200 \div 200$	
7.	12 halves ÷ 2 halves	
8.	$\dfrac{12}{2} \div \dfrac{2}{2}$	
9.	12 fourths ÷ 3 fourths	
10.	$\dfrac{12}{4} \div \dfrac{3}{4}$	
11.	$\dfrac{12}{8} \div \dfrac{3}{8}$	
12.	$\dfrac{2}{4} \div \dfrac{1}{4}$	
13.	$\dfrac{1}{4} \div \dfrac{2}{4}$	
14.	$\dfrac{4}{5} \div \dfrac{2}{5}$	
15.	$\dfrac{2}{5} \div \dfrac{4}{5}$	
16.	$\dfrac{3}{5} \div \dfrac{4}{5}$	
17.	$\dfrac{6}{8} \div \dfrac{2}{8}$	
18.	$\dfrac{6}{8} \div \dfrac{4}{8}$	
19.	$\dfrac{6}{8} \div \dfrac{5}{8}$	
20.	$\dfrac{6}{10} \div \dfrac{2}{10}$	
21.	$\dfrac{7}{10} \div \dfrac{8}{10}$	
22.	$\dfrac{4}{10} \div \dfrac{7}{10}$	

23.	$\dfrac{6}{12} \div \dfrac{4}{12}$	
24.	$\dfrac{6}{12} \div \dfrac{2}{6} = \dfrac{6}{12} \div \dfrac{\ }{12}$	
25.	$\dfrac{8}{14} \div \dfrac{7}{14}$	
26.	$\dfrac{8}{14} \div \dfrac{1}{2} = \dfrac{8}{14} \div \dfrac{\ }{14}$	
27.	$\dfrac{11}{14} \div \dfrac{2}{14}$	
28.	$\dfrac{11}{14} \div \dfrac{1}{7} = \dfrac{11}{14} \div \dfrac{\ }{14}$	
29.	$\dfrac{1}{7} \div \dfrac{6}{14} = \dfrac{\ }{14} \div \dfrac{6}{14}$	
30.	$\dfrac{7}{18} \div \dfrac{3}{18}$	
31.	$\dfrac{7}{18} \div \dfrac{1}{6} = \dfrac{7}{18} \div \dfrac{\ }{18}$	
32.	$\dfrac{1}{3} \div \dfrac{12}{18} = \dfrac{\ }{18} \div \dfrac{12}{18}$	
33.	$\dfrac{1}{6} \div \dfrac{4}{18}$	
34.	$\dfrac{4}{12} \div \dfrac{8}{6}$	
35.	$\dfrac{1}{3} \div \dfrac{3}{15}$	
36.	$\dfrac{2}{6} \div \dfrac{1}{9} = \dfrac{\ }{18} \div \dfrac{\ }{18}$	
37.	$\dfrac{1}{6} \div \dfrac{4}{9}$	
38.	$\dfrac{2}{3} \div \dfrac{3}{4}$	
39.	$\dfrac{1}{3} \div \dfrac{3}{5}$	
40.	$\dfrac{1}{7} \div \dfrac{1}{2}$	
41.	$\dfrac{5}{6} \div \dfrac{2}{9}$	
42.	$\dfrac{5}{9} \div \dfrac{2}{6}$	
43.	$\dfrac{5}{6} \div \dfrac{4}{9}$	
44.	$\dfrac{1}{2} \div \dfrac{4}{5}$	

Lesson 2:   The Relationship of Multiplication and Division

EUREKA
MATH™

Division of Fractions—Round 2 [KEY]

Directions: Evaluate each expression and simplify. Use blank spaces to create like units, where applicable.

1.	12 ones ÷ 2 ones	$\frac{12}{2} = 6$	23.	$\frac{6}{12} \div \frac{4}{12}$	$\frac{6}{4} = 1\frac{1}{2}$
2.	12 ÷ 2	$\frac{12}{2} = 6$	24.	$\frac{6}{12} \div \frac{2}{6} = \frac{6}{12} \div \frac{\ }{12}$	$\frac{6}{4} = 1\frac{1}{2}$
3.	12 tens ÷ 2 tens	$\frac{12}{2} = 6$	25.	$\frac{8}{14} \div \frac{7}{14}$	$\frac{8}{7} = 1\frac{1}{7}$
4.	120 ÷ 20	$\frac{12}{2} = 6$	26.	$\frac{8}{14} \div \frac{1}{2} = \frac{8}{14} \div \frac{\ }{14}$	$\frac{8}{7} = 1\frac{1}{7}$
5.	12 hundreds ÷ 2 hundreds	$\frac{12}{2} = 6$	27.	$\frac{11}{14} \div \frac{2}{14}$	$\frac{11}{2} = 5\frac{1}{2}$
6.	1,200 ÷ 200	$\frac{12}{2} = 6$	28.	$\frac{11}{14} \div \frac{1}{7} = \frac{11}{14} \div \frac{\ }{14}$	$\frac{11}{2} = 5\frac{1}{2}$
7.	12 halves ÷ 2 halves	$\frac{12}{2} = 6$	29.	$\frac{1}{7} \div \frac{6}{14} = \frac{\ }{14} \div \frac{6}{14}$	$\frac{2}{6} = \frac{1}{3}$
8.	$\frac{12}{2} \div \frac{2}{2}$	$\frac{12}{2} = 6$	30.	$\frac{7}{18} \div \frac{3}{18}$	$\frac{7}{3} = 2\frac{1}{3}$
9.	12 fourths ÷ 3 fourths	$\frac{12}{3} = 4$	31.	$\frac{7}{18} \div \frac{1}{6} = \frac{7}{18} \div \frac{\ }{18}$	$\frac{7}{3} = 2\frac{1}{3}$
10.	$\frac{12}{4} \div \frac{3}{4}$	$\frac{12}{3} = 4$	32.	$\frac{1}{3} \div \frac{12}{18} = \frac{\ }{18} \div \frac{12}{18}$	$\frac{6}{12} = \frac{1}{2}$
11.	$\frac{12}{8} \div \frac{3}{8}$	$\frac{12}{3} = 4$	33.	$\frac{1}{6} \div \frac{4}{18}$	$\frac{3}{4}$
12.	$\frac{2}{4} \div \frac{1}{4}$	$\frac{2}{1} = 2$	34.	$\frac{4}{12} \div \frac{8}{6}$	$\frac{4}{16} = \frac{1}{4}$
13.	$\frac{1}{4} \div \frac{2}{4}$	$\frac{1}{2}$	35.	$\frac{1}{3} \div \frac{3}{15}$	$\frac{5}{3} = 1\frac{2}{3}$
14.	$\frac{4}{5} \div \frac{2}{5}$	$\frac{4}{2} = 2$	36.	$\frac{2}{6} \div \frac{1}{9} = \frac{\ }{18} \div \frac{\ }{18}$	$\frac{6}{2} = 3$
15.	$\frac{2}{5} \div \frac{4}{5}$	$\frac{2}{4} = \frac{1}{2}$	37.	$\frac{1}{6} \div \frac{4}{9}$	$\frac{3}{8}$
16.	$\frac{3}{5} \div \frac{4}{5}$	$\frac{3}{4}$	38.	$\frac{2}{3} \div \frac{3}{4}$	$\frac{8}{9}$
17.	$\frac{6}{8} \div \frac{2}{8}$	$\frac{6}{2} = 3$	39.	$\frac{1}{3} \div \frac{3}{5}$	$\frac{5}{9}$
18.	$\frac{6}{8} \div \frac{4}{8}$	$\frac{6}{4} = 1\frac{1}{2}$	40.	$\frac{1}{7} \div \frac{1}{2}$	$\frac{2}{7}$
19.	$\frac{6}{8} \div \frac{5}{8}$	$\frac{6}{5} = 1\frac{1}{5}$	41.	$\frac{5}{6} \div \frac{2}{9}$	$\frac{15}{4} = 3\frac{3}{4}$
20.	$\frac{6}{10} \div \frac{2}{10}$	$\frac{6}{2} = 3$	42.	$\frac{5}{9} \div \frac{2}{6}$	$\frac{10}{6} = 1\frac{2}{3}$
21.	$\frac{7}{10} \div \frac{8}{10}$	$\frac{7}{8}$	43.	$\frac{5}{6} \div \frac{4}{9}$	$\frac{15}{8} = 1\frac{7}{8}$
22.	$\frac{4}{10} \div \frac{7}{10}$	$\frac{4}{7}$	44.	$\frac{1}{2} \div \frac{4}{5}$	$\frac{5}{8}$

©2015 Great Minds. eureka-math.org
G6-M4-TE-B4-1.3.1-01.2016

Lesson 3: The Relationship of Multiplication and Addition

Student Outcomes

- Students build and clarify the relationship of multiplication and addition by evaluating identities such as $3 \cdot g = g + g + g$.

Lesson Notes

Students continue to use the squares from Lessons 1 and 2 to create tape diagrams. Each pair of students needs 30 squares to complete the activities.

Classwork

Opening Exercise (5 minutes)

> **Opening Exercise**
>
> Write two different expressions that can be depicted by the tape diagram shown. One expression should include addition, while the other should include multiplication.
>
> a.   ☐☐☐   ☐☐☐   ☐☐☐
>
>   *Possible answers:* $3 + 3 + 3$ *or* 3×3
>
> b.   ☐☐☐☐☐☐☐☐   ☐☐☐☐☐☐☐☐
>
>   *Possible answers:* $8 + 8$ *or* 2×8
>
> c.   ☐☐☐☐☐   ☐☐☐☐☐   ☐☐☐☐☐
>
>   *Possible answers:* $5 + 5 + 5$ *or* 3×5

Discussion (17 minutes)

Provide each pair of students with a collection of 30 squares, which they use to create tape diagrams throughout the lesson.

- One partner builds a tape diagram to represent the expression $2 + 2 + 2 + 2$, while the other partner builds a tape diagram to represent 4×2.

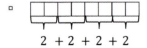

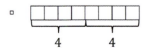

EUREKA MATH™

- What do you notice about the two tape diagrams you created?
  - *Possible answer: Although the tape diagrams represent two different expressions, they each have the same number of squares.*
- Why are the two tape diagrams the same? What does it say about the value of the expressions?
  - *The two tape diagrams are the same because the values of the expressions are equivalent.*
- If both expressions yield the same value, is there an advantage to using one over the other?
  - *Answers will vary.*
- Since each tape diagram has the same number of squares, can we say the two expressions are equivalent? Why or why not?
  - *Possible answer: The two expressions are equivalent because they represent the same value. When evaluated, both expressions will equal 8.*
- Therefore, $2 + 2 + 2 + 2 = 4 \times 2$. Let's build a new set of tape diagrams. One partner builds a tape diagram to represent the expression 3×4, while the other partner builds a tape diagram to represent the expression $4 + 4 + 4$.

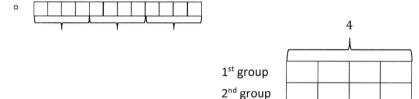

- Is 3×4 equivalent to $4 + 4 + 4$? Why or why not?
  - *Possible answer: The two expressions are equivalent because when each of them is evaluated, they equal 12, as we can see with our tape diagrams.*
- Using variables, write an equation to show the relationship of multiplication and addition.

Provide students with time to create an equation.

  - *Possible answer: $3g = g + g + g$. Emphasize that each g represents the same number.*

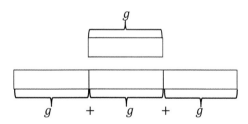

- $3g$ is the same as writing $3 \times g$, but we no longer use the $\times$ for multiplication because it looks like a variable and can become confusing. When a number is next to a variable with no sign, multiplication is implied.
- In the two previous lessons, we talked about identities. Is the equation $3g = g + g + g$ also an identity? Why or why not?
  - *Possible answer: The equation $3g = g + g + g$ is an identity because we can replace g with any number, and the equation will always be true.*

MP.2

Exercises (15 minutes)

Students can continue to work with the given squares and with their partners to answer the following questions.

Exercises

1. Write the addition sentence that describes the model and the multiplication sentence that describes the model.

$5 + 5 + 5$ *and* 3×5

2. Write an equivalent expression to demonstrate the relationship of multiplication and addition.

    a. $6 + 6$

    2×6

    b. $3 + 3 + 3 + 3 + 3 + 3$

    6×3

    c. $4 + 4 + 4 + 4 + 4$

    5×4

    d. 6×2

    $2 + 2 + 2 + 2 + 2 + 2$

    e. 4×6

    $6 + 6 + 6 + 6$

    f. 3×9

    $9 + 9 + 9$

    g. $h + h + h + h + h$

    $5h$

    h. $6y$

    $y + y + y + y + y + y$

EUREKA
MATH™

3. Roberto is not familiar with tape diagrams and believes that he can show the relationship of multiplication and addition on a number line. Help Roberto demonstrate that the expression 3×2 is equivalent to $2 + 2 + 2$ on a number line.

Possible answer: The first number line shows that there are 3 groups of 2, resulting in 6. The second number line shows the sum of $2 + 2 + 2$, resulting in 6.

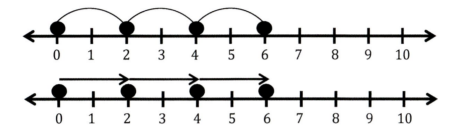

Since both number lines start at 0 and end at 6, the expressions are equivalent.

4. Tell whether the following equations are true or false. Then, explain your reasoning.

   a. $x + 6g - 6g = x$

   The equation is true because it demonstrates the addition identity.

   b. $2f - 4e + 4e = 2f$

   The equation is true because it demonstrates the subtraction identity.

5. Write an equivalent expression to demonstrate the relationship between addition and multiplication.

   a. $6 + 6 + 6 + 4 + 4 + 4$

   $4 \times 6 + 3 \times 4$

   b. $d + d + d + w + w + w + w + w$

   $3d + 5w$

   c. $a + a + b + b + b + c + c + c + c$

   $2a + 3b + 4c$

©2015 Great Minds. eureka-math.org
G6-M4-TE-B4-1.3.1-01.2016

Closing (4 minutes)

- Create a diagram that models 3 groups of size b.

    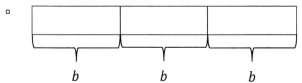

- Write two equivalent expressions that represent this model.

    - *Possible answers:* $3b, b + b + b$

- Peter says that since the addition expression yields the same value as the multiplication expression, he will always choose to use the addition expression when solving these types of problems. Convince Peter that he may want to reconsider his position.

    - *Answers will vary but should include the idea that when the group size is large, it is more advantageous to multiply instead of add.*

Exit Ticket (4 minutes)

©2015 Great Minds. eureka-math.org
G6-M4-TE-B4-1.3.1-01.2016

Name _____   Date _____

Lesson 3: The Relationship of Multiplication and Addition

Exit Ticket

Write an equivalent expression to show the relationship of multiplication and addition.

1. $8 + 8 + 8 + 8 + 8 + 8 + 8 + 8 + 8$

2. 4×9

3. $6 + 6 + 6$

4. $7h$

5. $j + j + j + j + j$

6. $u + u + u + u + u + u + u + u + u + u$

Exit Ticket Sample Solutions

Write an equivalent expression to show the relationship of multiplication and addition.

1. $8 + 8 + 8 + 8 + 8 + 8 + 8 + 8 + 8$

     9×8

2. 4×9

     $9 + 9 + 9 + 9$

3. $6 + 6 + 6$

     3×6

4. $7h$

     $h + h + h + h + h + h + h$

5. $j + j + j + j + j$

     $5j$

6. $u + u + u + u + u + u + u + u + u + u$

     $10u$

Problem Set Sample Solutions

Write an equivalent expression to show the relationship of multiplication and addition.

1. $10 + 10 + 10$

     3×10

2. $4 + 4 + 4 + 4 + 4 + 4 + 4$

     7×4

3. 8×2

     $2 + 2 + 2 + 2 + 2 + 2 + 2 + 2$

4. 3×9

     $9 + 9 + 9$

5. $6m$

     $m + m + m + m + m + m$

6. $d + d + d + d + d$

     $5d$

EUREKA
MATH™

Lesson 4: The Relationship of Division and Subtraction

Student Outcomes

- Students build and clarify the relationship of division and subtraction by determining that $12 \div x = 4$ means $12 - x - x - x - x = 0$.

Lesson Notes

Students continue to use the squares from Lessons 1–3 to create tape diagrams. Each pair of students needs 30 squares to complete the activities.

Classwork

Discussion (20 minutes)

Provide each pair of students with a collection of 30 squares so they can use these squares to create tape diagrams throughout the lesson.

- Build a tape diagram that has 20 squares.

  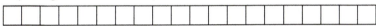

- Divide the tape diagram into 4 equal sections.

- How many squares are in each of the 4 sections?

  - 5

- Write a number sentence to demonstrate what happened.

  - $20 \div 4 = 5$

- Combine your squares again to have a tape diagram with 20 squares.

  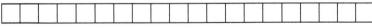

- Now, subtract 4 squares from your tape diagram.

  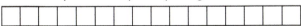

- Write an expression to demonstrate what happened.

  - $20 - 4$

- Subtract 4 more squares, and alter your expression to represent the new tape diagram.

  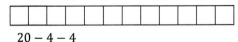

  - $20 - 4 - 4$

- Subtract 4 more squares, and alter your expression to represent the new tape diagram.

  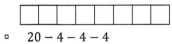

  - $20 - 4 - 4 - 4$

- Subtract 4 more squares, and alter your expression to represent the new tape diagram.

    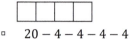

    □ $20 - 4 - 4 - 4 - 4$

- Last time. Subtract 4 more squares, and alter your expression to an equation in order to represent a number sentence showing the complete transformation of the tape diagram.

    □ *No squares should remain.*

    □ $20 - 4 - 4 - 4 - 4 - 4 = 0$

- Let's take a look at the process we took to determine the difference to be zero.

Discuss the process step-by-step to determine that the number of times the divisor was subtracted from the dividend is the same number as the quotient.

$$20 - 4 = 16$$
$$16 - 4 = 12$$
$$12 - 4 = 8$$
$$8 - 4 = 4$$
$$4 - 4 = 0$$

Subtracted 5 times.
Quotient is 5.

- Do you recognize a relationship between $20 \div 4 = 5$ and $20 - 4 - 4 - 4 - 4 - 4 = 0$? If so, what is it?

    □ *Possible answer: If you subtract the divisor from the dividend 5 times (the quotient), there will be no remaining squares.*

- Let's take a look at a similar problem, $20 \div 5 = 4$, to see if the quotient is the number of times the divisor is subtracted from the dividend.

- Let's create a number sentence when we subtract the divisor.

    MP.8

    □ $20 - 5 - 5 - 5 - 5 = 0$

Discuss the process to determine that the number of times the divisor is subtracted from the dividend is the same number as the quotient.

$$20 - 5 = 15$$
$$15 - 5 = 10$$
$$10 - 5 = 5$$
$$5 - 5 = 0$$

Subtracted 4 times.
Quotient is 4.

- Determine the relationship between $20 \div 5 = 4$ and $20 - 5 - 5 - 5 - 5 = 0$.

    MP.8

    □ $20 \div 5 = 4$ *can be interpreted as subtracting* 4 *fives from* 20 *is* 0, *or* $20 - 5 - 5 - 5 - 5 = 0$.

- Is this relationship always true? Let's try to prove that it is.

Model the following set of tape diagrams with leading questions for discussion.

- x is a number. What does $20 \div x = 5$ mean?

    □ *Exactly five x's can be subtracted from twenty.*

- What must x be in this division sentence?

    □ 4

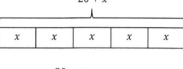

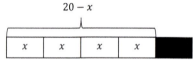

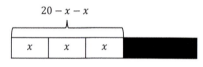

EUREKA MATH™

- Let's keep taking x away until we reach zero.

Model taking each x away and creating subtraction expressions to record.

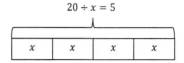

- Build a subtraction expression.
  - $20 - x - x - x - x - x = 0$
- Is $20 - 4 - 4 - 4 - 4 - 4 = 0$?
  - *Yes*

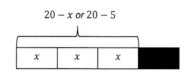

MP.2

- Develop two equations using numbers and letters to show the relationship of division and subtraction.
  - *Possible answers:* $20 \div x = 5$ *and*
    $20 - x - x - x - x - x = 0$

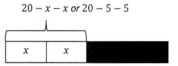

- Or $20 \div x = 5$ means that 5 can be subtracted exactly x number of times from 20. Is it true when $x = 4$?
- To determine if $x = 4$, let's keep taking x away until we reach zero.

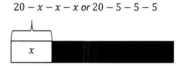

Model taking each x away and creating subtraction expressions to record by following the diagram to the right.

- Build a subtraction equation.
  - $20 - 5 - 5 - 5 - 5 = 0$
- What two operations are we relating in the problems we completed?
  - *Division and subtraction*

Exercise 1 (10 minutes)

Students work in pairs to answer the following questions.

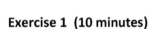

Exercise 1

Build subtraction equations using the indicated equations. The first example has been completed for you.

Division Equation	Divisor Indicates the Size of the Unit	Tape Diagram	What is x, y, z?
$12 \div x = 4$	$12 - x - x - x - x = 0$	$12 - 3 - 3 - 3 - 3 = 0; x = 3$ units in each group	$x = 3$
$18 \div x = 3$	$18 - x - x - x = 0$	$18 - 6 - 6 - 6 = 0; x = 6$ units in each group	$x = 6$
$35 \div y = 5$	$35 - y - y - y - y - y = 0$	$35 - 7 - 7 - 7 - 7 - 7 = 0; y = 7$ units in each group	$y = 7$
$42 \div z = 6$	$42 - z - z - z - z - z - z = 0$	$42 - 7 - 7 - 7 - 7 - 7 - 7 = 0; z = 7$ units in each group	$z = 7$

Division Equation	Divisor Indicates the Number of Units	Tape Diagram	What is x, y, z?
$12 \div x = 4$	$12 - 4 - 4 - 4 = 0$	$12 - 4 - 4 - 4 = 0; x = 3$ groups	$x = 3$
$18 \div x = 3$	$18 - 3 - 3 - 3 - 3 - 3 - 3 = 0$	$18 - 3 - 3 - 3 - 3 - 3 - 3 = 0; x = 6$ groups	$x = 6$
$35 \div y = 5$	$35 - 5 - 5 - 5 - 5 - 5 - 5 - 5 = 0$	$35 - 5 - 5 - 5 - 5 - 5 - 5 - 5 = 0; y = 7$ groups	$y = 7$
$42 \div z = 6$	$42 - 6 - 6 - 6 - 6 - 6 - 6 - 6 = 0$	$42 - 6 - 6 - 6 - 6 - 6 - 6 - 6 = 0; z = 7$ groups	$z = 7$

Exercise 2 (5 minutes)

Exercise 2

Answer each question using what you have learned about the relationship of division and subtraction.

a. If $12 \div x = 3$, how many times would x have to be subtracted from 12 in order for the answer to be zero? What is the value of x?

$3; x = 4$

b. $36 - f - f - f - f = 0$. Write a division sentence for this repeated subtraction sentence. What is the value of f?

$36 \div 4 = f$ or $36 \div f = 4; f = 9$

c. If $24 \div b = 12$, which number is being subtracted 12 times in order for the answer to be zero?

Two

EUREKA
MATH

Closing (5 minutes)

Display the graphic organizer provided at the end of the lesson. Copies of the organizer can be made for students to follow along and record.

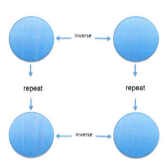

- In each of the circles, we can place an operation to satisfy the organizer. In the last four lessons, we have discovered that each operation has a relationship with other operations, whether they are inverse operations or they are repeats of another.

Place the addition symbol in the upper left-hand circle.

- Let's start with addition. What is the inverse operation of addition?
  - *Subtraction*

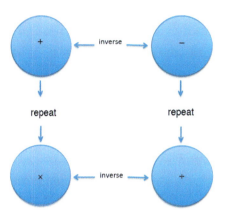

Place the subtraction symbol in the upper right-hand circle.

- After our discussion today, repeated subtraction can be represented by which operation?
  - *Division*

Place the division symbol in the lower right-hand circle.

- Which operation is the inverse of division?
  - *Multiplication*

Place the multiplication symbol in the lower left-hand circle.

- Let's see if this is correct. Is multiplication the repeat operation of addition?
  - *Yes*

Understanding the relationships of operations is going to be instrumental when solving equations later in this unit.

Exit Ticket (5 minutes)

Name _____   Date _____

Lesson 4: The Relationship of Division and Subtraction

Exit Ticket

1. Represent $56 \div 8 = 7$ using subtraction. Explain your reasoning.

2. Explain why $30 \div x = 6$ is the same as $30 - x - x - x - x - x - x = 0$. What is the value of x in this example?

©2015 Great Minds. eureka-math.org
G6-M4-TE-B4-1.3.1-01.2016

Exit Ticket Sample Solutions

1. Represent $56 \div 8 = 7$ using subtraction. Explain your reasoning.

    $56 - 7 - 7 - 7 - 7 - 7 - 7 - 7 - 7 = 0$ *because*

    $56 - 7 = 49;\ 49 - 7 = 42;\ 42 - 7 = 35;\ 35 - 7 = 28;\ 28 - 7 = 21;\ 21 - 7 = 14;\ 14 - 7 = 7;\ 7 - 7 = 0.$

    OR

    $56 - 8 - 8 - 8 - 8 - 8 - 8 - 8 = 0$ *because*

    $56 - 8 = 48;\ 48 - 8 = 40;\ 40 - 8 = 32;\ 32 - 8 = 24;\ 24 - 8 = 16;\ 16 - 8 = 8;\ 8 - 8 = 0.$

2. Explain why $30 \div x = 6$ is the same as $30 - x - x - x - x - x = 0$. What is the value of x in this example?

    $30 \div 5 = 6$, *so* $x = 5$. *When I subtract* 5 *from* 30 *six times, the result is zero. Division is a repeat operation of subtraction.*

Problem Set Sample Solutions

Build subtraction equations using the indicated equations.

	Division Equation	Divisor Indicates the Size of the Unit	Tape Diagram	What is x, y, z?
1.	$24 \div x = 4$	$24 - x - x - x - x = 0$	$24 - x - x - x - x = 0; x = 6$ units in each group	$x = 6$
2.	$36 \div x = 6$	$36 - x - x - x - x - x - x = 0$	$36 - x - x - x - x - x = 0; x = 6$ units in each group	$x = 6$
3.	$28 \div y = 7$	$28 - y - y - y - y - y - y - y = 0$	$28 - y - y - y - y - y - y = 0; y = 4$ units in each group	$y = 4$
4.	$30 \div y = 5$	$30 - y - y - y - y - y = 0$	$30 - y - y - y - y = 0; y = 6$ units in each group	$y = 6$
5.	$16 \div z = 4$	$16 - z - z - z - z = 0$	$16 - z - z - z = 0; z = 4$ units in each group	$z = 4$

©2015 Great Minds. eureka-math.org
G6-M4-TE-B4-1.3.1-01.2016

	Division Equation	Divisor Indicates the Number of Units	Tape Diagram	What is x, y, z?
1.	$24 \div x = 4$	$24 - 4 - 4 - 4 - 4 - 4 - 4 = 0$	$24 - 4 - 4 - 4 - 4 - 4 - 4 = 0; x = 6\ groups$	$x = 6$
2.	$36 \div x = 6$	$36 - 6 - 6 - 6 - 6 - 6 - 6 = 0$	$36 - 6 - 6 - 6 - 6 - 6 = 0; x = 6\ groups$	$x = 6$
3.	$28 \div y = 7$	$28 - 7 - 7 - 7 - 7 = 0$	$28 - 7 - 7 - 7 - 7 = 0; y = 4\ groups$	$y = 4$
4.	$30 \div y = 5$	$30 - 5 - 5 - 5 - 5 - 5 - 5 = 0$	$30 - 5 - 5 - 5 - 5 - 5 - 5 = 0; y = 6\ groups$	$y = 6$
5.	$16 \div z = 4$	$16 - 4 - 4 - 4 - 4 = 0$	$16 - 4 - 4 - 4 - 4 = 0; z = 4\ groups$	$z = 4$

Lesson 4:   The Relationship of Division and Subtraction

EUREKA MATH™

Graphic Organizer Reproducible

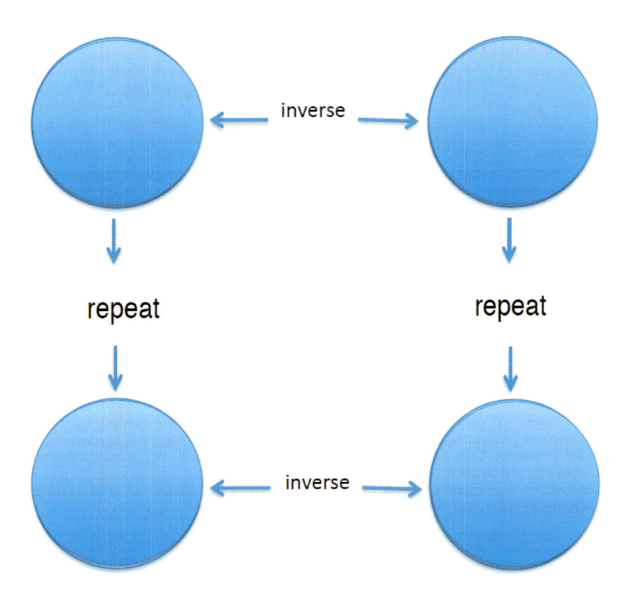

GRADE 6

Mathematics Curriculum

Topic B

Special Notations of Operations

6.EE.A.1, 6.EE.A.2c

Focus Standards:	6.EE.A.1	Write and evaluate numerical expressions involving whole-number exponents.
	6.EE.A.2c	Write, read, and evaluate expressions in which letters stand for numbers.
	c.	Evaluate expressions at specific values of their variables. Include expressions that arise from formulas used in real-world problems. Perform arithmetic operations, including those involving whole-number exponents, in the conventional order when there are no parentheses to specify a particular order (Order of Operations). *For example, use the formulas $V = s^3$ and $A = 6s^2$ to find the volume and surface area of a cube with sides of length $s = 1/2$.*
Instructional Days:	2	
Lesson 5:	Exponents (S)[1]	
Lesson 6:	The Order of Operations (P)	

In Topic B, students differentiate between the product of two numbers and whole numbers with exponents. They differentiate between the two through exploration of patterns, specifically noting how squares grow from a 1×1 measure. They determine that a square with a length and width of three units in measure is constructed with nine square units. This expression is represented as 3^2 and is evaluated as the product of $3 \times 3 = 9$, not the product of the base and exponent, 6. They further differentiate between the two by comparing the areas of two models with similar measures, as shown below.

$2a$
When $a = 3$ the area of the rectangle is 6 square units.

a^2
When $a = 3$ the area of the rectangle is 9 square units.

---

[1]Lesson Structure Key: **P**-Problem Set Lesson, **M**-Modeling Cycle Lesson, **E**-Exploration Lesson, **S**-Socratic Lesson

Once students understand that the base is multiplied by itself the number of times as stated by the exponent, they make a smooth transition into bases that are represented with positive fractions and decimals. They know that for any number a, a^m is defined as the product of m factors of a. The number a is the base, and m is called the *exponent* (or the *power*) of a.

In Lesson 6, students build on their previous understanding of the order of operations by including work with exponents. They follow the order of operations to evaluate numerical expressions. They recognize that, in the absence of parentheses, exponents are evaluated first. Students identify when the order of operations is incorrectly applied and determine the applicable course to correctly evaluate expressions. They understand that the placement of parentheses can alter the final solution when evaluating expressions, as in the following example:

$$2^4 \cdot (2 + 8) - 16 \qquad\qquad 2^4 \cdot 2 + 8 - 16$$
$$2^4 \cdot 10 - 16 \qquad\qquad 16 \cdot 2 + 8 - 16$$
$$16 \cdot 10 - 16 \qquad\qquad 32 + 8 - 16$$
$$160 - 16 \qquad\qquad 40 - 16$$
$$144 \qquad\qquad 24$$

Students continue to apply the order of operations throughout the module as they evaluate numerical and algebraic expressions.

©2015 Great Minds. eureka-math.org
G6-M4-TE-B4-1.3.1-01.2016

Lesson 5: Exponents

Student Outcomes

- Students discover that $3x = x + x + x$ is not the same thing as x^3, which is $x \cdot x \cdot x$.
- Students understand that a base number can be represented with a positive whole number, positive fraction, or positive decimal and that for any number a, a^m is defined as the product of m factors of a. The number a is the base, and m is called the *exponent* or *power* of a.

Lesson Notes

In Grade 5, students are introduced to exponents. Explain patterns in the number of zeros of the product when multiplying a number by powers of 10, and explain patterns in the placement of the decimal point when a decimal is multiplied or divided by a power of 10. Use whole number exponents to denote powers of 10 (**5.NBT.A.2**).

In this lesson, students use new terminology (*base, squared,* and *cubed*) and practice moving between exponential notation, expanded notation, and standard notation. The following terms should be displayed, defined, and emphasized throughout Lesson 5: *base, exponent, power, squared,* and *cubed*.

Classwork

Fluency Exercise (5 minutes): Multiplication of Decimals

RWBE: Refer to the Rapid White Board Exchanges section in the Module Overview for directions on how to administer an RWBE.

Opening Exercise (2 minutes)

> **Opening Exercise**
>
> As you evaluate these expressions, pay attention to how you arrive at your answers.
>
> $4 + 4 + 4 + 4 + 4 + 4 + 4 + 4 + 4 + 4$
>
> $9 + 9 + 9 + 9 + 9$
>
> $10 + 10 + 10 + 10 + 10$

Discussion (15 minutes)

- How many of you solved the problems by *counting on*? That is, starting with 4, you counted on 4 more each time $(5, 6, 7, \mathbf{8}, 9, 10, 11, \mathbf{12}, 13, 14, 15, \mathbf{16}, \dots)$.
- If you did not find the answer that way, could you have done so?
  - *Yes, but it is time consuming and cumbersome.*
- Addition is a faster way of counting on.

 EUREKA MATH™

- How else could you find the sums using addition?

    ▫ *Count by 4, 9, or 10.*

- How else could you solve the problems?

    ▫ *Multiply 4 times 10; multiply 9 times 5; or multiply 10 times 5.*

- Multiplication is a faster way to add numbers when the addends are the same.

- When we add five groups of 10, we use an abbreviation and a different notation, called *multiplication*. $10 + 10 + 10 + 10 + 10 = 5 \times 10$

- If multiplication is a more efficient way to represent addition problems involving the repeated addition of the same addend, do you think there might be a more efficient way to represent the repeated multiplication of the same factor, as in $10 \times 10 \times 10 \times 10 \times 10 = ?$

MP.2 & MP.7

Allow students to make suggestions; some may recall this from previous lessons.

$$10 \times 10 \times 10 \times 10 \times 10 = 10^5$$

- We see that when we add five groups of 10, we write 5×10, but when we multiply five copies of 10, we write 10^5. So, multiplication by 5 in the context of addition corresponds exactly to the exponent 5 in the context of multiplication.

Make students aware of the correspondence between addition and multiplication because what they know about *repeated addition* helps them learn exponents as *repeated multiplication* going forward.

- The little 5 we write is called an *exponent* and is written as a *superscript*. The numeral 5 is written only half as tall and half as wide as the 10, and the bottom of the 5 should be halfway up the number 10. The top of the 5 can extend a little higher than the top of the zero in 10. Why do you think we write exponents so carefully?

    ▫ *It reduces the chance that a reader will confuse 10^5 with 105.*

> **Scaffolding:**
>
> When teaching students how to write an exponent as a *superscript*, compare and contrast the notation with how to write a *subscript*, as in the molecular formula for water, H_2O, or carbon dioxide, CO_2. Here the number is again half as tall as the capital letters, and the top of the 2 is halfway down it. The bottom of the subscript can extend a little lower than the bottom of the letter. Ignore the meaning of a chemical subscript.

Examples 1–5 (5 minutes)

Work through Examples 1–5 as a group; supplement with additional examples if needed.

> **Examples 1–10**
>
> Write each expression in exponential form.
>
> 1. $5 \times 5 \times 5 \times 5 \times 5 = 5^5$ 2. $2 \times 2 \times 2 \times 2 = 2^4$
>
> Write each expression in expanded form.
>
> 3. $8^3 = 8 \times 8 \times 8$ 4. $10^6 = 10 \times 10 \times 10 \times 10 \times 10 \times 10$
>
> 5. $g^3 = g \times g \times g$

- The repeated factor is called the *base,* and the exponent is also called the *power.* Say the numbers in Examples 1–5 to a partner.

Check to make sure students read the examples correctly:

□   *Five to the fifth power, two to the fourth power, eight to the third power, ten to the sixth power, and g to the third power.*

Go back to Examples 1–4, and use a calculator to evaluate the expressions.

1. $5 \times 5 \times 5 \times 5 \times 5 = 5^5 = 3,125$ 2. $2 \times 2 \times 2 \times 2 = 2^4 = 16$

3. $8^3 = 8 \times 8 \times 8 = 512$ 4. $10^6 = 10 \times 10 \times 10 \times 10 \times 10 \times 10 = 1,000,000$

What is the difference between $3g$ and g^3?

$3g = g + g + g$ or 3 times g; $g^3 = g \times g \times g$

Take time to clarify this important distinction.

■   The base number can also be written in decimal or fraction form. Try Examples 6, 7, and 8. Use a calculator to evaluate the expressions.

Examples 6–8 (4 minutes)

---

6.   **Write the expression in expanded form, and then evaluate.**

   $(3.8)^4 = 3.8 \times 3.8 \times 3.8 \times 3.8 = 208.5136$

7.   **Write the expression in exponential form, and then evaluate.**

   $2.1 \times 2.1 = (2.1)^2 = 4.41$

8.   **Write the expression in exponential form, and then evaluate.**

   $0.75 \times 0.75 \times 0.75 = (0.75)^3 = 0.421875$

Note to teacher:

If students need additional help multiplying fractions, refer to the first four modules of Grade 5.

The base number can also be a fraction. Convert the decimals to fractions in Examples 7 and 8 and evaluate. Leave your answer as a fraction. Remember how to multiply fractions!

Example 7:

$\frac{21}{10} \times \frac{21}{10} = \left(\frac{21}{10}\right)^2 = \frac{441}{100} = 4\frac{41}{100}$

Example 8:

$\frac{3}{4} \times \frac{3}{4} \times \frac{3}{4} = \left(\frac{3}{4}\right)^3 = \frac{27}{64}$

EUREKA
MATH™

Examples 9–10 (1 minute)

---

9. Write the expression in exponential form, and then evaluate.

$$\frac{1}{2} \times \frac{1}{2} \times \frac{1}{2} = \left(\frac{1}{2}\right)^3 = \frac{1}{8}$$

10. Write the expression in expanded form, and then evaluate.

$$\left(\frac{2}{3}\right)^2 = \frac{2}{3} \times \frac{2}{3} = \frac{4}{9}$$

- There is a special name for numbers raised to the second power. When a number is raised to the second power, it is called *squared*. Remember that in geometry, squares have the same two dimensions: length and width. For $b > 0$, b^2 is the area of a square with side length b.

- What is the value of 5 squared?
    - 25

- What is the value of 7 squared?
    - 49

- What is the value of 8 squared?
    - 64

- What is the value of 1 squared?
    - 1

A multiplication chart is included at the end of this lesson. Post or project it as needed.

- Where are square numbers found on the multiplication table?
    - *On the diagonal*

- There is also a special name for numbers raised to the third power. When a number is raised to the third power, it is called *cubed*. Remember that in geometry, cubes have the same three dimensions: length, width, and height. For $b > 0$, b^3 is the volume of a cube with edge length b.

- What is the value of 1 cubed?
    - $1 \times 1 \times 1 = 1$

- What is the value of 2 cubed?
    - $2 \times 2 \times 2 = 8$

- What is the value of 3 cubed?
    - $3 \times 3 \times 3 = 27$

- In general, for any number $x, x^1 = x$, and for any positive integer $n > 1$, x^n is, by definition,

$$x^n = \underbrace{(x \cdot x \cdots x)}_{n \text{ times}}.$$

MP.6

- What does the x represent in this equation?
    - *The x represents the factor that will be repeatedly multiplied by itself.*

- What does the n represent in this expression?
    - *n represents the number of times x will be multiplied.*

- Let's look at this with some numbers. How would we represent 4^n?
  - $4^n = \underbrace{(4 \cdot 4 \cdots 4)}_{n \text{ times}}$
- What does the 4 represent in this expression?
  - *The 4 represents the factor that will be repeatedly multiplied by itself.*
- What does the n represent in this expression?
  - *n represents the number of times 4 will be multiplied.*

MP.6

- What if we were simply multiplying? How would we represent $4n$?
  - *Because multiplication is repeated addition, $4n = \underbrace{(4 + 4 \cdots 4)}_{n \text{ times}}$.*
- What does the 4 represent in this expression?
  - *The 4 represents the addend that will be repeatedly added to itself.*
- What does the n represent in this expression?
  - *n represents the number of times 4 will be added.*

Exercises (8 minutes)

Ask students to fill in the chart, supplying the missing expressions.

Exercises

1. Fill in the missing expressions for each row. For whole number and decimal bases, use a calculator to find the standard form of the number. For fraction bases, leave your answer as a fraction.

Exponential Form	Expanded Form	Standard Form
3^2	3×3	9
2^6	$2 \times 2 \times 2 \times 2 \times 2 \times 2$	64
4^5	$4 \times 4 \times 4 \times 4 \times 4$	1,024
$\left(\frac{3}{4}\right)^2$	$\frac{3}{4} \times \frac{3}{4}$	$\frac{9}{16}$
$(1.5)^2$	1.5×1.5	2.25

2. Write five cubed in all three forms: exponential form, expanded form, and standard form.

   $5^3; 5 \times 5 \times 5; 125$

3. Write fourteen and seven-tenths squared in all three forms.

   $(14.7)^2; 14.7 \times 14.7; 216.09$

4. One student thought two to the third power was equal to six. What mistake do you think he made, and how would you help him fix his mistake?

   The student multiplied the base, 2, by the exponent, 3. This is wrong because the exponent never multiplies the base; the exponent tells how many copies of the base are to be used as factors.

Closing (2 minutes)

- We use multiplication as a quicker way to do repeated addition if the addends are the same. We use exponents as a quicker way to multiply if the factors are the same.

- Carefully write exponents as superscripts to avoid confusion.

Lesson Summary

EXPONENTIAL NOTATION FOR WHOLE NUMBER EXPONENTS: Let m be a nonzero whole number. For any number a, the expression a^m is the product of m factors of a, i.e.,

$$a^m = \underbrace{a \cdot a \cdots \cdot a}_{m \text{ times}}.$$

The number a is called the *base,* and m is called the *exponent* or *power* of a.

When m is 1, "the product of one factor of a" just means a (i.e., $a^1 = a$). Raising any nonzero number a to the power of 0 is defined to be 1 (i.e., $a^0 = 1$ for all $a \neq 0$).

Exit Ticket (3 minutes)

Name _____   Date _____

Lesson 5: Exponents

Exit Ticket

1. What is the difference between $6z$ and z^6?

2. Write 10^3 as a multiplication expression having repeated factors.

3. Write $8 \times 8 \times 8 \times 8$ using an exponent.

Exit Ticket Sample Solutions

1. What is the difference between $6z$ and z^6?

   $6z = z + z + z + z + z + z$ or 6 times z; $z^6 = z \times z \times z \times z \times z \times z$

2. Write 10^3 as a multiplication expression having repeated factors.

   $10 \times 10 \times 10$

3. Write $8 \times 8 \times 8 \times 8$ using an exponent.

   8^4

Problem Set Sample Solutions

1. Complete the table by filling in the blank cells. Use a calculator when needed.

Exponential Form	Expanded Form	Standard Form
3^5	$3 \times 3 \times 3 \times 3 \times 3$	243
4^3	$4 \times 4 \times 4$	64
$(1.9)^2$	1.9×1.9	3.61
$\left(\dfrac{1}{2}\right)^5$	$\dfrac{1}{2} \times \dfrac{1}{2} \times \dfrac{1}{2} \times \dfrac{1}{2} \times \dfrac{1}{2}$	$\dfrac{1}{32}$

2. Why do whole numbers raised to an exponent get greater, while fractions raised to an exponent get smaller?

   As whole numbers are multiplied by themselves, products are larger because there are more groups. As fractions of fractions are taken, the product is smaller. A part of a part is less than how much we started with.

3. The powers of 2 that are in the range 2 through $1,000$ are 2, 4, 8, 16, 32, 64, 128, 256, and 512. Find all the powers of 3 that are in the range 3 through $1,000$.

   $3, 9, 27, 81, 243, 729$

4. Find all the powers of 4 in the range 4 through $1,000$.

   $4, 16, 64, 256$

5. Write an equivalent expression for $n \times a$ using only addition.

   $\underbrace{(a + a + \cdots a)}_{n\ times}$

6. Write an equivalent expression for w^b using only multiplication.

$$w^b = \underbrace{(w \cdot w \cdot \,\cdots\, \cdot w)}_{b \ times}$$

    a.   Explain what w is in this new expression.

        w is the factor that will be repeatedly multiplied by itself.

    b.   Explain what b is in this new expression.

        b is the number of times w will be multiplied.

7. What is the advantage of using exponential notation?

    It is a shorthand way of writing a multiplication expression if the factors are all the same.

8. What is the difference between $4x$ and x^4? Evaluate both of these expressions when $x = 2$.

    $4x$ means four times x; this is the same as $x + x + x + x$. On the other hand, x^4 means x to the fourth power, or $x \times x \times x \times x$.

    When $x = 2$, $4x = 4 \times 2 = 8$.

    When $x = 2$, $x^4 = 2 \times 2 \times 2 \times 2 = 16$.

Multiplication of Decimals

Progression of Exercises

1. $0.5 \times 0.5 =$

   0.25

2. $0.6 \times 0.6 =$

   0.36

3. $0.7 \times 0.7 =$

   0.49

4. $0.5 \times 0.6 =$

   0.3

5. $1.5 \times 1.5 =$

   2.25

6. $2.5 \times 2.5 =$

   6.25

7. $0.25 \times 0.25 =$

   0.0625

8. $0.1 \times 0.1 =$

   0.01

9. $0.1 \times 123.4 =$

   12.34

10. $0.01 \times 123.4 =$

    1.234

Lesson 6: The Order of Operations

Student Outcomes

- Students evaluate numerical expressions. They recognize that in the absence of parentheses, exponents are evaluated first.

Classwork

Opening (5 minutes)

Take a few minutes to review the Problem Set from the previous lesson. Clarify any misconceptions about the use and evaluation of exponents.

Opening Exercise (5 minutes)

Post the following expression on the board, and ask students to evaluate it.

$$3 + 4 \times 2$$

Ask students to record their answers and report them using personal white boards, cards, or electronic vote devices. Students who arrive at an answer other than 11 or 14 should recheck their work.

Discussion (5 minutes)

- How did you evaluate the expression $3 + 4 \times 2$?
  - *I added $3 + 4$ first for a sum of 7; then, I multiplied 7×2 for a product of 14.*
  - *I multiplied 4×2 first for a product of 8; then, I added $8 + 3$ for a sum of 8.*
- Only one of these answers can be correct. When we evaluate expressions, we must agree to use one set of rules so that everyone arrives at the same correct answer.
- During the last lesson, we said that addition was a shortcut to *counting on*. How could you think about subtraction?
  - *Subtraction is a shortcut to "counting back."*
- These were the first operations that you learned because they are the least complicated. Next, you learned about multiplication and division.
- Multiplication can be thought of as repeated addition. Thinking back on Lesson 4, how could you think about division?
  - *Division is repeated subtraction.*
- Multiplication and division are more powerful than addition and subtraction, which led mathematicians to develop the order of operations in this way. When we evaluate expressions that have any of these four operations, we always calculate multiplication and division before doing any addition or subtraction. Since multiplication and division are equally powerful, we simply evaluate these two operations as they are written in the expression, from left to right.

©2015 Great Minds. eureka-math.org
G6-M4-TE-B4-1.3.1-01.2016

- Addition and subtraction are at the same level in the order of operations and are evaluated from left to right in an expression. Now that these rules of order of operations are clear, can you go back and evaluate the expression $3 + 4 \times 2$ as 11?

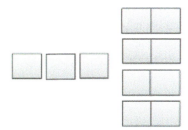

- The diagram correctly models the expression $3 + 4 \times 2$.
- With addition, we are finding the sum of two addends. In this example, the first addend is the number 3. The second addend happens to be the number that is the value of the expression 4×2; so, before we can add, we must determine the value of the second addend.

Example 1 (5 minutes): Expressions with Only Addition, Subtraction, Multiplication, and Division

> **Example 1: Expressions with Only Addition, Subtraction, Multiplication, and Division**
>
> **What operations are evaluated first?**
>
> *Multiplication and division are evaluated first, from left to right.*
>
> **What operations are always evaluated last?**
>
> *Addition and subtraction are always evaluated last, from left to right.*

Ask students to evaluate the expressions.

> **Exercises 1–3**
>
> 1. $4 + 2 \times 7$
>
> $4 + 14$
>
> 18
>
> 2. $36 \div 3 \times 4$
>
> 12×4
>
> 48
>
> 3. $20 - 5 \times 2$
>
> $20 - 10$
>
> 10

- In the last lesson, you learned about exponents, which are a way of writing repeated multiplication. So, exponents are more powerful than multiplication or division. If exponents are present in an expression, they are evaluated before any multiplication or division.

- We now know that when we evaluate expressions, we must agree to use one set of rules so that everyone arrives at the same correct answer. These rules are based on doing the most powerful operations first (exponents), then the less powerful ones (multiplication and division, going from left to right), and finally, the least powerful ones last (addition and subtraction, going from left to right).

- Evaluate the expression $4 + 6 \times 6 \div 8$.

    □ $4 + (6 \times 6) \div 8$

        $4 + (36 \div 8)$

        $4 + 4.5$

        8.5

- Now, evaluate the expression $4 + 6^2 \div 8$.

    □ $4 + (6^2) \div 8$

        $4 + (36 \div 8)$

        $4 + 4.5$

        8.5

- Why was your first step to find the value of 6^2?

    □   *Because exponents are evaluated first.*

Example 2 (5 minutes): Expressions with Four Operations and Exponents

Display the following expression.

> **Example 2: Expressions with Four Operations and Exponents**
>
> $$4 + 9^2 \div 3 \times 2 - 2$$
>
> **What operation is evaluated first?**
>
> *Exponents ($9^2 = 9 \times 9 = 81$)*
>
> **What operations are evaluated next?**
>
> *Multiplication and division, from left to right ($81 \div 3 = 27$; $27 \times 2 = 54$)*
>
> **What operations are always evaluated last?**
>
> *Addition and subtraction, from left to right ($4 + 54 = 58$; $58 - 2 = 56$)*
>
> **What is the final answer?**
>
> *56*

Scaffolding:

Some students may benefit from rewriting the expression on successive lines, evaluating only one or two operations on each line.

EUREKA MATH

- Evaluate the next two exercises.

While the answers are provided, it is extremely important to circulate to ensure that students are using the correct order of operations to achieve the answer. For example, in Exercise 5, they should show 4^3 first, followed by 2×8.

Exercises 4–5

4. $90 - 5^2 \times 3$

   $90 - 25 \times 3$

   $90 - 75$

   15

5. $4^3 + 2 \times 8$

   $64 + 2 \times 8$

   $64 + 16$

   80

Example 3 (5 minutes): Expressions with Parentheses

- The last important rule in the order of operations involves grouping symbols, usually parentheses. These tell us that in certain circumstances or scenarios, we need to do things out of the usual order. Operations inside grouping symbols are always evaluated first, before exponents and any operations.

Example 3: Expressions with Parentheses

Consider a family of 4 that goes to a soccer game. Tickets are $\$5.00$ each. The mom also buys a soft drink for $\$2.00$. How would you write this expression?

$4 \times 5 + 2$

How much will this outing cost?

$\$22$

- Here is a model of the scenario:

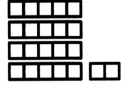

Consider a different scenario: The same family goes to the game as before, but each of the family members wants a drink. How would you write this expression?

$4 \times (5 + 2)$

> **Why would you add the 5 and 2 first?**
>
> *We need to determine how much each person spends. Each person spends $7; then, we multiply by 4 people to figure out the total cost.*
>
> **How much will this outing cost?**
>
> $28

- Here is a model of the second scenario:

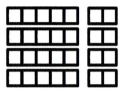

> **How many groups are there?**
>
> 4
>
> **What does each group comprise?**
>
> $5 + $2, *or* $7

- The last complication that can arise is if two or more sets of parentheses are ever needed; evaluate the innermost parentheses first, and then work outward.
- Try Exercises 6 and 7.

> **Exercises 6–7**
>
> 6. $2 + (9^2 - 4)$
>
> $2 + (81 - 4)$
>
> $2 + 77$
>
> 79
>
>
> 7. $2 \cdot \big(13 + 5 - 14 \div (3 + 4)\big)$
>
> $2 \cdot (13 + 5 - 14 \div 7)$
>
> $2 \cdot (13 + 5 - 2)$
>
> $2 \cdot 16$
>
> 32

If students are confused trying to divide 14 by 3, reiterate the rule about nested parentheses.

EUREKA
MATH™

Example 4 (5 minutes): Expressions with Parentheses and Exponents

▪ Let's take a look at how parentheses and exponents work together. Sometimes a problem will have parentheses, and the values inside the parentheses have an exponent. Let's evaluate the following expression.

Place the expression on the board.

▪ We will evaluate the parentheses first.

Example 4: Expressions with Parentheses and Exponents

$$2 \times (3 + 4^2)$$

Which value will we evaluate first within the parentheses? Evaluate.

First, evaluate 4^2, which is 16; then, add 3. The value of the parentheses is 19.

$2 \times (3 + 4^2)$

$2 \times (3 + 16)$

2×19

Evaluate the rest of the expression.

$2 \times 19 = 38$

Place the expression on the board:

What do you think will happen when the exponent in this expression is outside of the parentheses?

$$2 \times (3 + 4)^2$$

Will the answer be the same?

Answers will vary.

Which should we evaluate first? Evaluate.

Parentheses

$2 \times (3 + 4)^2$

$2 \times (7)^2$

What happened differently here than in our last example?

The 4 was not raised to the second power because it did not have an exponent. We simply added the values inside the parentheses.

What should our next step be?

We need to evaluate the exponent next.

$7^2 = 7 \times 7 = 49$

Evaluate to find the final answer.

2×49

98

What do you notice about the two answers?

The final answers were not the same.

What was different between the two expressions?

Answers may vary. In the first problem, a value inside the parentheses had an exponent, and that value was evaluated first because it was inside of the parentheses. In the second problem, the exponent was outside of the parentheses, which made us evaluate what was in the parentheses first; then, we raised that value to the power of the exponent.

What conclusions can you draw about evaluating expressions with parentheses and exponents?

Answers may vary. Regardless of the location of the exponent in the expression, evaluate the parentheses first. Sometimes there will be values with exponents inside the parentheses. If the exponent is outside the parentheses, evaluate the parentheses first, and then evaluate to the power of the exponent.

- Try Exercises 8 and 9.

Exercises 8–9

8. $7 + (12 - 3^2)$

$7 + (12 - 9)$

$7 + 3$

10

9. $7 + (12 - 3)^2$

$7 + 9^2$

$7 + 81$

88

EUREKA
MATH

Closing (5 minutes)

- When we evaluate expressions, we use one set of rules so that everyone arrives at the same correct answer. Grouping symbols, like parentheses, tell us to evaluate whatever is inside them before moving on. These rules are based on doing the most powerful operations first (exponents), then the less powerful ones (multiplication and division, going from left to right), and finally, the least powerful ones last (addition and subtraction, going from left to right).

Lesson Summary

NUMERICAL EXPRESSION: A *numerical expression* is a number, or it is any combination of sums, differences, products, or divisions of numbers that evaluates to a number.

Statements like "3 +" or "3 ÷ 0" are not numerical expressions because neither represents a point on the number line. Note: Raising numbers to whole number powers are considered numerical expressions as well since the operation is just an abbreviated form of multiplication, e.g., $2^3 = 2 \cdot 2 \cdot 2$.

VALUE OF A NUMERICAL EXPRESSION: The *value of a numerical expression* is the number found by evaluating the expression.

For example: $\frac{1}{3} \cdot (2 + 4) + 7$ is a numerical expression, and its value is 9.

Note: Please do not stress words over meaning here. It is okay to talk about the *number computed, computation, calculation,* and so on to refer to the value as well.

Exit Ticket (5 minutes)

Name _____   Date _____

Lesson 6: The Order of Operations

Exit Ticket

1. Evaluate this expression: $39 \div (2 + 1) - 2 \times (4 + 1)$.

2. Evaluate this expression: $12 \times (3 + 2^2) \div 2 - 10$.

3. Evaluate this expression: $12 \times (3 + 2)^2 \div 2 - 10$.

Exit Ticket Sample Solutions

1. **Evaluate this expression:** $39 \div (2 + 1) - 2 \times (4 + 1)$.

    $39 \div 3 - 2 \times 5$

    $13 - 10$

    3

2. **Evaluate this expression:** $12 \times (3 + 2^2) \div 2 - 10$.

    $12 \times (3 + 4) \div 2 - 10$

    $12 \times 7 \div 2 - 10$

    $84 \div 2 - 10$

    $42 - 10$

    32

3. **Evaluate this expression:** $12 \times (3 + 2)^2 \div 2 - 10$.

    $12 \times 5^2 \div 2 - 10$

    $12 \times 25 \div 2 - 10$

    $300 \div 2 - 10$

    $150 - 10$

    140

Problem Set Sample Solutions

Evaluate each expression.

1. $3 \times 5 + 2 \times 8 + 2$

    $15 + 16 + 2$

    33

2. $(\$1.75 + 2 \times \$0.25 + 5 \times \$0.05) \times 24$

    $(\$1.75 + \$0.50 + \$0.25) \times 24$

    $\$2.50 \times 24$

    $\$60.00$

3. $(2 \times 6) + (8 \times 4) + 1$

    $12 + 32 + 1$

    45

4. $((8 \times 1.95) + (3 \times 2.95) + 10.95) \times 1.06$

   $(15.6 + 8.85 + 10.95) \times 1.06$

   35.4×1.06

   37.524

5. $((12 \div 3)^2 - (18 \div 3^2)) \times (4 \div 2)$

   $(4^2 - (18 \div 9)) \times (4 \div 2)$

   $(16 - 2) \times 2$

   14×2

   28

Mathematics Curriculum

Topic C

Replacing Letters and Numbers

6.EE.A.2c, 6.EE.A.4

Focus Standards:	6.EE.A.2c	Write, read, and evaluate expressions in which letters stand for numbers.
		c. Evaluate expressions at specific values of their variables. Include expressions that arise from formulas used in real-world problems. Perform arithmetic operations, including those involving whole-number exponents, in the conventional order when there are no parentheses to specify a particular order (Order of Operations). *For example, use the formulas $V = s^3$ and $A = 6s^2$ to find the volume and surface area of a cube with sides of length $s = 1/2$.*
	6.EE.A.4	Identify when two expressions are equivalent (i.e., when the two expressions name the same number regardless of which value is substituted into them). *For example, the expressions $y + y + y$ and $3y$ are equivalent because they name the same number regardless of which number y stands for.*
Instructional Days:	2	
Lesson 7:	Replacing Letters with Numbers (P)[1]	
Lesson 8:	Replacing Numbers with Letters (S)	

Students begin substituting, or replacing, letters with numbers and numbers with letters in Topic C in order to evaluate expressions with a given number and to determine expressions to create identities. In Lesson 7, students replace letters with a given number in order to evaluate the expression to one number.

---

[1]Lesson Structure Key: **P**-Problem Set Lesson, **M**-Modeling Cycle Lesson, **E**-Exploration Lesson, **S**-Socratic Lesson

They continue to practice with exponents in this lesson in order to determine the area of squares and rectangles as shown below.

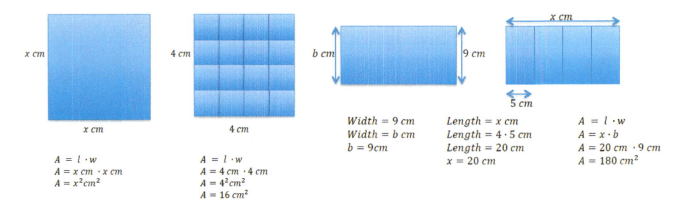

$A = l \cdot w$
$A = x\ cm \cdot x\ cm$
$A = x^2 cm^2$

$A = l \cdot w$
$A = 4\ cm \cdot 4\ cm$
$A = 4^2 cm^2$
$A = 16\ cm^2$

$Width = 9\ cm$
$Width = b\ cm$
$b = 9cm$

$Length = x\ cm$
$Length = 4 \cdot 5\ cm$
$Length = 20\ cm$
$x = 20\ cm$

$A = l \cdot w$
$A = x \cdot b$
$A = 20\ cm \cdot 9\ cm$
$A = 180\ cm^2$

In Lesson 8, students understand that a number in an expression can be replaced with a letter to determine identities. Through replacement of numbers, students discover and build identities such as $a + b = b + a$, $a \times b = b \times a$, $g \times 1 = g$, $g + 0 = g$, $g \div 1 = g$, $g \div g = 1$, $1 \div g = \frac{1}{g}$. These identities aid in solving equations with variables, as well as problem solving with equations.

$$4 \times 1 = 4 \qquad\qquad g \times 1 = g$$
$$4 \div 1 = 4 \qquad\qquad g \div 1 = g$$
$$4 \times 0 = 0 \qquad\qquad g \times 0 = 0$$
$$1 \div 4 = \frac{1}{4} \qquad\qquad 1 \div g = \frac{1}{g}$$

$$3 + 4 = 4 + 3 \qquad\qquad a + 4 = 4 + a$$
$$3 \times 4 = 4 \times 3 \qquad\qquad a \times 4 = 4 \times a$$
$$3 + 3 + 3 + 3 = 4 \times 3 \qquad\qquad a + a + a + a = 4 \times a$$
$$3 \div 4 = \frac{3}{4} \qquad\qquad a \div 4 = \frac{a}{4}$$

EUREKA
MATH™

Lesson 7: Replacing Letters with Numbers

Student Outcomes

- Students understand that a letter represents one number in an expression. When that number replaces the letter, the expression can be evaluated to one number.

Lesson Notes

Before this lesson, make it clear to students that, just like 3×3 is 3^2 or three squared, units $\times$ units is units^2 or units squared (also called *square units*).

It may be helpful to cut and paste some of the figures from this lesson onto either paper or an interactive white board application. Each of the basic figures is depicted two ways: One has side lengths that can be counted, and the other is a similar figure without grid lines. Also, ahead of time, draw a 23 cm square on a chalkboard, a white board, or an interactive board.

There is a square in the student materials that is approximately 23 mm square, or 529 mm^2.

Classwork

Example 1 (10 minutes)

Draw or project the square shown.

Example 1

What is the length of one side of this square?

3 units

What is the formula for the area of a square?

$A = s^2$

What is the square's area as a multiplication expression?

3 units $\times$ *3 units*

What is the square's area?

9 *square units*

We can count the units. However, look at this other square. Its side length is 23 cm. That is just too many tiny units to draw. What expression can we build to find this square's area?

23 cm

23 cm × 23 cm

What is the area of the square? Use a calculator if you need to.

529 cm²

- A letter represents *one* number in an expression. That number was 3 in our first square and 23 in our second square. When that number replaces the letter, the expression can be evaluated to *one* number. In our first example, the expression was evaluated to be 9, and in the second example, the expression was evaluated to be 529.

Make sure students understand that 9 is one number, but 529 is also one number. (It happens to have 3 digits, but it is still one number.)

Exercise 1 (5 minutes)

Ask students to work both problems from Exercise 1 in their student materials. Make clear to students that these drawings are not to scale.

Exercise 1

Complete the table below for both squares. Note: These drawings are not to scale.

$s = 4$

$s = 25$ in.

EUREKA
MATH

Length of One Side of the Square	Square's Area Written as an Expression	Square's Area Written as a Number
4 *units*	4 *units* × 4 *units*	16 *square units*
25 in.	25 in.× 25 in.	625 in²

Make sure students have the units correctly recorded in each of the cells of the table. When units are not specified, keep the label *unit* or *square unit*.

Example 2 (10 minutes)

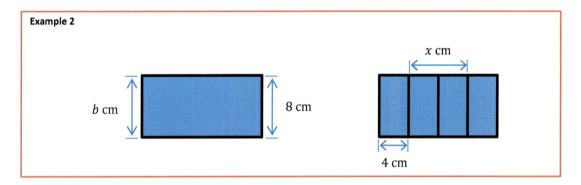

Example 2

- The formula $A = l \times w$ is an efficient way to find the area of a rectangle without being required to count the area units in a rectangle.

> **What does the letter *b* represent in this blue rectangle?**
>
> $b = 8$

Give students a short time for discussion of the next question among partners, and then ask for an answer and an explanation.

> **With a partner, answer the following question: Given that the second rectangle is divided into four *equal* parts, what number does the x represent?**
>
> $x = 8$
>
> **How did you arrive at this answer?**
>
> *We reasoned that each width of the 4 congruent rectangles must be the same. Two 4 cm lengths equals 8 cm.*
>
> **What is the total length of the second rectangle? Tell a partner how you know.**
>
> *The length consists of 4 segments that each has a length of 4 cm. 4 × 4 cm = 16 cm.*

> If the two large rectangles have equal lengths and widths, find the area of each rectangle.
>
> $8 \text{ cm} \times 16 \text{ cm} = 128 \text{ cm}^2$
>
> Discuss with your partner how the formulas for the area of squares and rectangles can be used to evaluate area for a particular figure.

- Remember, a letter represents *one* number in an expression. When that number replaces the letter, the expression can be evaluated to *one* number.

Exercise 2 (5 minutes)

Ask students to complete the table for both rectangles in their student materials. Using a calculator is appropriate.

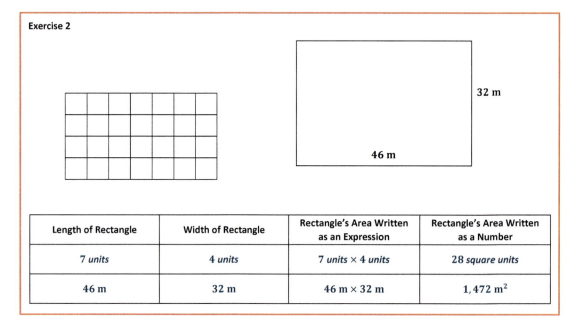

Exercise 2

Length of Rectangle	Width of Rectangle	Rectangle's Area Written as an Expression	Rectangle's Area Written as a Number
7 *units*	4 *units*	7 *units* × 4 *units*	28 *square units*
46 m	32 m	46 m × 32 m	1,472 m²

EUREKA
MATH™

Example 3 (3 minutes)

- The formula $V = l \times w \times h$ is a quick way to determine the volume of right rectangular prisms.
- Take a look at the right rectangular prisms in your student materials.

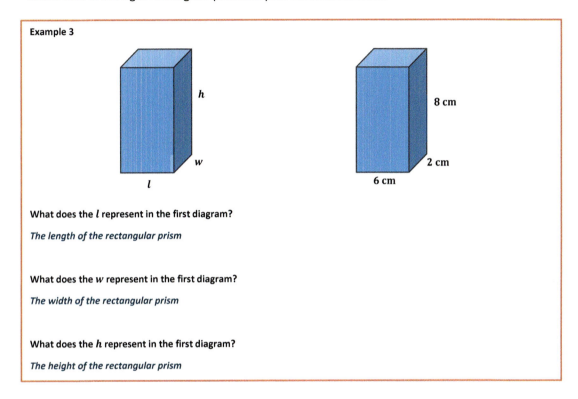

Example 3

What does the l represent in the first diagram?

The length of the rectangular prism

What does the w represent in the first diagram?

The width of the rectangular prism

What does the h represent in the first diagram?

The height of the rectangular prism

- Notice that the right rectangular prism in the second diagram is an exact copy of the first diagram.

Since we know the formula to find the volume is $V = l \times w \times h$, what number can we substitute for the l in the formula? Why?

6, because the length of the second right rectangular prism is 6 cm.

What other number can we substitute for the l?

No other number can replace the l. Only one number can replace one letter.

What number can we substitute for the w in the formula? Why?

2, because the width of the second right rectangular prism is 2 cm.

What number can we substitute for the h in the formula?

8, because the height of the second right rectangular prism is 8 cm.

©2015 Great Minds. eureka-math.org
G6-M4-TE-B4-1.3.1-01.2016

> Determine the volume of the second right rectangular prism by replacing the letters in the formula with their appropriate numbers.
>
> $V = l \times w \times h$; $V = 6 \text{ cm} \times 2 \text{ cm} \times 8 \text{ cm} = 96 \text{ cm}^3$

Exercise 3 (5 minutes)

Ask students to complete the table for both figures in their student materials. Using a calculator is appropriate.

Exercise 3

Complete the table for both figures. Using a calculator is appropriate.

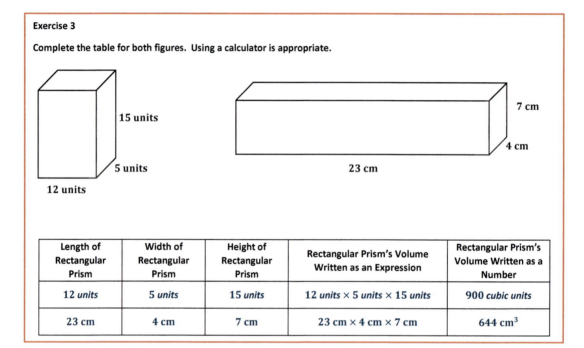

Length of Rectangular Prism	Width of Rectangular Prism	Height of Rectangular Prism	Rectangular Prism's Volume Written as an Expression	Rectangular Prism's Volume Written as a Number
12 *units*	5 *units*	15 *units*	12 *units* × 5 *units* × 15 *units*	900 *cubic units*
23 cm	4 cm	7 cm	23 cm × 4 cm × 7 cm	644 cm³

EUREKA
MATH™

Closing (2 minutes)

- How many numbers are represented by one letter in an expression?
    - *One*
- When that number replaces the letter, the expression can be evaluated to what?
    - *One number*

Lesson Summary

VARIABLE (description): A *variable* is a symbol (such as a letter) that is a placeholder for a number.

EXPRESSION (description): An *expression* is a numerical expression, or it is the result of replacing some (or all) of the numbers in a numerical expression with variables.

There are two ways to build expressions:

1. We can start out with a numerical expression, such as $\frac{1}{3} \cdot (2 + 4) + 7$, and replace some of the numbers with letters to get $\frac{1}{3} \cdot (x + y) + z$.

2. We can build such expressions from scratch, as in $x + x(y - z)$, and note that if numbers were placed in the expression for the variables x, y, and z, the result would be a numerical expression.

The key is to strongly link expressions back to computations with numbers.

The description for *expression* given above is meant to work nicely with how students in Grade 6 and Grade 7 learn to manipulate expressions. In these grades, a lot of time is spent *building expressions* and *evaluating expressions*. Building and evaluating helps students see that expressions are really just a slight abstraction of arithmetic in elementary school. Building often occurs by thinking about examples of numerical expressions first and then replacing the numbers with letters in a numerical expression. The act of evaluating for students at this stage means they replace each of the variables with specific numbers and then compute to obtain a number.

Exit Ticket (5 minutes)

Name _____   Date _____

Lesson 7: Replacing Letters with Numbers

Exit Ticket

1. In the drawing below, what do the letters l and w represent?

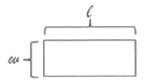

2. What does the expression $l + w + l + w$ represent?

3. What does the expression $l \cdot w$ represent?

4. The rectangle below is congruent to the rectangle shown in Problem 1. Use this information to evaluate the expressions from Problems 2 and 3.

EUREKA
MATH™

Exit Ticket Sample Solutions

1. In the drawing below, what do the letters l and w represent?

   Length and width of the rectangle

2. What does the expression $l + w + l + w$ represent?

   Perimeter of the rectangle, or the sum of the sides of the rectangle

3. What does the expression $l \cdot w$ represent?

   Area of the rectangle

4. The rectangle below is congruent to the rectangle shown in Problem 1. Use this information to evaluate the expressions from Problems 2 and 3.

   $l = 5$ *and* $w = 2$ $P = 14$ *units* $A = 10$ *units²*

Problem Set Sample Solutions

1. Replace the side length of this square with 4 in., and find the area.

   s

   The student should draw a square, label the side 4 in., and calculate the area to be 16 *in².*

2. Complete the table for each of the given figures.

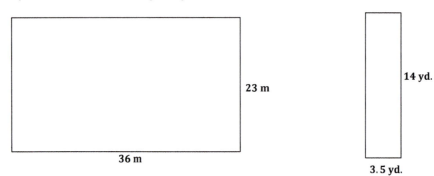

23 m

36 m

14 yd.

3.5 yd.

Length of Rectangle	Width of Rectangle	Rectangle's Area Written as an Expression	Rectangle's Area Written as a Number
36 m	23 m	$36 \text{ m} \times 23 \text{ m}$	828 m^2
14 yd.	3.5 yd.	$14 \text{ yd.} \times 3.5 \text{ yd.}$	49 yd^2

3. Find the perimeter of each quadrilateral in Problems 1 and 2.

    $P = 16$ in. $P = 118$ m $P = 35$ yd.

4. Using the formula $V = l \times w \times h$, find the volume of a right rectangular prism when the length of the prism is 45 cm, the width is 12 cm, and the height is 10 cm.

    $V = l \times w \times h$; $V = 45 \text{ cm} \times 12 \text{ cm} \times 10 \text{ cm} = 5,400 \text{ cm}^3$

EUREKA
MATH™

Lesson 8: Replacing Numbers with Letters

Student Outcomes

- Students understand that a letter in an expression or an equation can represent a number. When that number is replaced with a letter, an expression or an equation is stated.
- Students discover the commutative properties of addition and multiplication, the additive identity property of zero, and the multiplicative identity property of one. They determine that $g \div 1 = g$, $g \div g = 1$, and $1 \div g = \dfrac{1}{g}$.

Classwork

Fluency Exercise (10 minutes): Division of Fractions II

Sprint: Refer to the Sprints and the Sprint Delivery Script sections in the Module Overview for directions on how to administer a Sprint.

Opening Exercise (5 minutes)

Write this series of equations on the board:

Opening Exercise

$$4 + 0 = 4$$

$$4 \times 1 = 4$$

$$4 \div 1 = 4$$

$$4 \times 0 = 0$$

$$1 \div 4 = \frac{1}{4}$$

Discussion (5 minutes)

MP.3

How many of these statements are true?

All of them

How many of those statements would be true if the number 4 was replaced with the number 7 in each of the number sentences?

All of them

Would the number sentences be true if we were to replace the number 4 with any other number?

©2015 Great Minds. eureka-math.org
G6-M4-TE-B4-1.3.1-01.2016

- Let students make conjectures about substitutions.

> **What if we replaced the number 4 with the number 0? Would each of the number sentences be true?**
>
> *No. The first four are true, but the last one, dividing by zero, is not true.*

MP.3

- Division by zero is *undefined*. You cannot make zero groups of objects, and group size cannot be zero.
- It appears that we can replace the number 4 with any nonzero number, and each of the number sentences will be true.
- A letter in an expression can represent a number. When that number is replaced with a letter, an expression is stated.

> **What if we replace the number 4 with a letter g? Please write all 4 expressions below, replacing each 4 with a g.**
>
> $$g + 0 = g$$
> $$g \times 1 = g$$
> $$g \div 1 = g$$
> $$g \times 0 = 0$$
> $$1 \div g = \frac{1}{g}$$
>
> **Are these all true (except for $g = 0$) when dividing?**
>
> *Yes*

- Let's look at each of these a little closer and see if we can make some generalizations.

Example 1 (5 minutes): Additive Identity Property of Zero

> **Example 1: Additive Identity Property of Zero**
>
> $$g + 0 = g$$
>
> **Remember a letter in a mathematical expression represents a number. Can we replace g with any number?**
>
> *Yes*
>
> **Choose a value for g, and replace g with that number in the equation. What do you observe?**
>
> *The value of g does not change when 0 is added to g.*
>
> **Repeat this process several times, each time choosing a different number for g.**

Allow students to experiment for about a minute. Most quickly realize the additive identity property of zero: Any number added to zero equals itself. The number's identity does not change.

Will all values of g result in a true number sentence?

Yes

Write the mathematical language for this property below:

$g + 0 = g$, *additive identity property of zero. Any number added to zero equals itself.*

Example 2 (5 minutes): Multiplicative Identity Property of One

---

Example 2: Multiplicative Identity Property of One

$$g \times 1 = g$$

Remember a letter in a mathematical expression represents a number. Can we replace g with any number?

Yes

Choose a value for g, and replace g with that number in the equation. What do you observe?

The value of g does not change when g is multiplied by 1.

Allow students to experiment for about a minute with the next question. Most quickly realize the multiplicative identity property of one: Any number multiplied by 1 equals itself. The number's identity does not change.

Will all values of g result in a true number sentence? Experiment with different values before making your claim.

Yes

Write the mathematical language for this property below:

$g \times 1 = g$, *multiplicative identity property of one. Any number multiplied by one equals itself.*

Example 3 (6 minutes): Commutative Property of Addition and Multiplication

---

Example 3: Commutative Property of Addition and Multiplication

$$3 + 4 = 4 + 3$$
$$3 \times 4 = 4 \times 3$$

Replace the 3's in these number sentences with the letter a.

$$a + 4 = 4 + a$$
$$a \times 4 = 4 \times a$$

> Choose a value for a, and replace a with that number in each of the equations. What do you observe?
>
> *The result is a true number sentence.*

Allow students to experiment for about a minute with the next question. Most quickly realize that the equations are examples of the commutative property of addition and commutative property of multiplication. These are sometimes called the "any-order properties."

> Will all values of a result in a true number sentence? Experiment with different values before making your claim.
>
> *Yes, any number, even zero, can be used in place of the variable a.*
>
> Now, write the equations again, this time replacing the number 4 with a variable, b.
>
> $$a + b = b + a$$
> $$a \times b = b \times a$$
>
> Will all values of a and b result in true number sentences for the first two equations? Experiment with different values before making your claim.
>
> *Yes*
>
> Write the mathematical language for this property below:
>
> $a + b = b + a$, *commutative property of addition. Order does not matter when adding.*
>
> $a \times b = b \times a$, *commutative property of multiplication. Order does not matter when multiplying.*

▪ Models are useful for making abstract ideas more concrete.

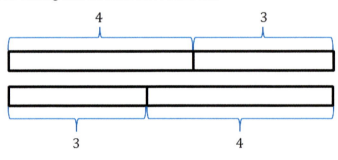

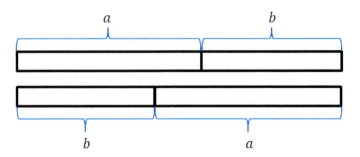

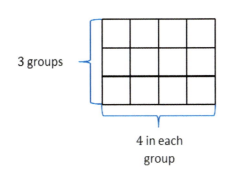

3 groups

4 in each group

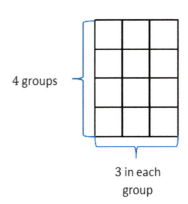

4 groups

3 in each group

Example 4 (4 minutes)

Display and discuss the models above as they relate to the commutative property of addition and the commutative property of multiplication. When finished, pose a new question:

- Will all values of a and b result in a true number sentence for the equation $a + a + a + a = b \times a$? Experiment with different values before making your claim.

Allow students to experiment for about a minute. They should discover that any value can be substituted for the variable a, but only 4 can be used for b, since there are exactly 4 copies of a in the equation.

- Summarize your discoveries with a partner.
  - *In the equation $a + a + a + a = b \times a$, any value can be substituted for the variable a, but only 4 can be used for b, since there are exactly 4 copies of a in the equation.*

- Finally, consider the last equation, $a \div b = \dfrac{a}{b}$. Is this true for all values of a and b?
  - *It is true for all values of a and all values of $b \neq 0$.*

Example 4

$$3 + 3 + 3 + 3 = 4 \times 3$$

$$3 \div 4 = \frac{3}{4}$$

Replace the 3's in these number sentences with the letter a.

$$a + a + a + a = 4 \times a$$

$$a \div 4 = \frac{a}{4}$$

Choose a value for a, and replace a with that number in each of the equations. What do you observe?

The result is a true number sentence.

Will all values of a result in a true number sentence? Experiment with different values before making your claim.

Yes, any number, even zero, can be used in place of the variable a.

Now, write the equations again, this time replacing the number 4 with a variable, b.

$$a + a + a + a = b \times a$$

$$a \div b = \frac{a}{b}, b \neq 0$$

Will all values of a and b result in true number sentences for the equations? Experiment with different values before making your claim.

In the equation $a + a + a + a = b \times a$, any value can be substituted for the variable a, but only 4 can be used for b since there are exactly 4 copies of a in the equation.

It is true for all values of a and all values of $b \neq 0$.

Closing (2 minutes)

- Tell your partner which of these properties of numbers is the easiest for you to remember.

Allow sharing for a short time.

- Now, tell your partner which of these properties of numbers is the hardest for you to remember.

Allow sharing for a short time.

- Although these properties might seem simple, we apply them in many different ways in mathematics. If you have a good grasp on them, you will recognize them and use them in many applications.

- With a partner, create two different division problems that support the following: $g \times 1 = g$, and be ready to explain your reasoning.

  □ *$g \div g = 1$; $5 \div 5 = 1$; $34 \div 34 = 1$; $2\frac{7}{8} \div 2\frac{7}{8} = 1$; and so on.*

  □ *Any nonzero number divided by itself equals 1.*

  □ *If a number g is divided into g equal parts, each part will have a size equal to one.*

  □ *If g items are divided into groups of size g, there will be one group.*

- What about any number divided by 1? What does this mean?

  □ *$g \div 1 = g$*

  □ *If a number g is divided into 1 part, then the size of that part will be g.*

  □ *Or, if g items are divided into 1 group, there will be g items in that group.*

Exit Ticket (5 minutes)

Name _____   Date _____

Lesson 8: Replacing Numbers with Letters

Exit Ticket

1. State the commutative property of addition, and provide an example using two different numbers.

2. State the commutative property of multiplication, and provide an example using two different numbers.

3. State the additive property of zero, and provide an example using any other number.

4. State the multiplicative identity property of one, and provide an example using any other number.

Exit Ticket Sample Solutions

1. State the commutative property of addition, and provide an example using two different numbers.

    Any two different addends can be chosen, such as $5 + 6 = 6 + 5$.

2. State the commutative property of multiplication, and provide an example using two different numbers.

    Any two different factors can be chosen, such as $4 \times 9 = 9 \times 4$.

3. State the additive property of zero, and provide an example using any other number.

    Any nonzero addend can be chosen, such as $3 + 0 = 3$.

4. State the multiplicative identity property of one, and provide an example using any other number.

    Any nonzero factor can be chosen, such as $12 \times 1 = 12$.

Problem Set Sample Solutions

1. State the commutative property of addition using the variables a and b.

    $a + b = b + a$

2. State the commutative property of multiplication using the variables a and b.

    $a \times b = b \times a$

3. State the additive property of zero using the variable b.

    $b + 0 = b$

4. State the multiplicative identity property of one using the variable b.

    $b \times 1 = b$

5. Demonstrate the property listed in the first column by filling in the third column of the table.

Commutative Property of Addition	$25 + c =$	$c + 25$
Commutative Property of Multiplication	$l \times w =$	$w \times l$
Additive Property of Zero	$h + 0 =$	h
Multiplicative Identity Property of One	$v \times 1 =$	v

6. Why is there no commutative property for subtraction or division? Show examples.

    Answers will vary. Examples should show reasoning and proof that the commutative property does not work for subtraction and division. An example would be $8 \div 2$ and $2 \div 8$. $8 \div 2 = 4$, but $2 \div 8 = \frac{1}{4}$.

Number Correct: _____

Division of Fractions II—Round 1

Directions: Determine the quotient of the fractions and simplify.

1.	$\dfrac{4}{10} \div \dfrac{2}{10}$	
2.	$\dfrac{9}{12} \div \dfrac{3}{12}$	
3.	$\dfrac{6}{10} \div \dfrac{4}{10}$	
4.	$\dfrac{2}{8} \div \dfrac{3}{8}$	
5.	$\dfrac{2}{7} \div \dfrac{6}{7}$	
6.	$\dfrac{11}{9} \div \dfrac{8}{9}$	
7.	$\dfrac{5}{13} \div \dfrac{10}{13}$	
8.	$\dfrac{7}{8} \div \dfrac{13}{16}$	
9.	$\dfrac{3}{5} \div \dfrac{7}{10}$	
10.	$\dfrac{9}{30} \div \dfrac{3}{5}$	
11.	$\dfrac{1}{3} \div \dfrac{4}{5}$	
12.	$\dfrac{2}{5} \div \dfrac{3}{4}$	
13.	$\dfrac{3}{4} \div \dfrac{5}{9}$	
14.	$\dfrac{4}{5} \div \dfrac{7}{12}$	
15.	$\dfrac{3}{8} \div \dfrac{5}{2}$	

16.	$3\dfrac{1}{8} \div \dfrac{2}{3}$	
17.	$1\dfrac{5}{6} \div \dfrac{1}{2}$	
18.	$\dfrac{5}{8} \div 2\dfrac{3}{4}$	
19.	$\dfrac{1}{3} \div 1\dfrac{4}{5}$	
20.	$\dfrac{3}{4} \div 2\dfrac{3}{10}$	
21.	$2\dfrac{1}{5} \div 1\dfrac{1}{6}$	
22.	$2\dfrac{4}{9} \div 1\dfrac{3}{5}$	
23.	$1\dfrac{2}{9} \div 3\dfrac{2}{5}$	
24.	$2\dfrac{2}{3} \div 3$	
25.	$1\dfrac{3}{4} \div 2\dfrac{2}{5}$	
26.	$4 \div 1\dfrac{2}{9}$	
27.	$3\dfrac{1}{5} \div 6$	
28.	$2\dfrac{5}{6} \div 1\dfrac{1}{3}$	
29.	$10\dfrac{2}{3} \div 8$	
30.	$15 \div 2\dfrac{3}{5}$	

Division of Fractions II—Round 1 [KEY]

Directions: Determine the quotient of the fractions and simplify.

1.	$\dfrac{4}{10} \div \dfrac{2}{10}$	$\dfrac{4}{2} = 2$	16.	$3\dfrac{1}{8} \div \dfrac{2}{3}$	$\dfrac{75}{16} = 4\dfrac{11}{16}$
2.	$\dfrac{9}{12} \div \dfrac{3}{12}$	$\dfrac{9}{3} = 3$	17.	$1\dfrac{5}{6} \div \dfrac{1}{2}$	$\dfrac{22}{6} = \dfrac{11}{3} = 3\dfrac{2}{3}$
3.	$\dfrac{6}{10} \div \dfrac{4}{10}$	$\dfrac{6}{4} = \dfrac{3}{2} = 1\dfrac{1}{2}$	18.	$\dfrac{5}{8} \div 2\dfrac{3}{4}$	$\dfrac{20}{88} = \dfrac{5}{22}$
4.	$\dfrac{2}{8} \div \dfrac{3}{8}$	$\dfrac{2}{3}$	19.	$\dfrac{1}{3} \div 1\dfrac{4}{5}$	$\dfrac{5}{27}$
5.	$\dfrac{2}{7} \div \dfrac{6}{7}$	$\dfrac{2}{6} = \dfrac{1}{3}$	20.	$\dfrac{3}{4} \div 2\dfrac{3}{10}$	$\dfrac{30}{92} = \dfrac{15}{46}$
6.	$\dfrac{11}{9} \div \dfrac{8}{9}$	$\dfrac{11}{8} = 1\dfrac{3}{8}$	21.	$2\dfrac{1}{5} \div 1\dfrac{1}{6}$	$\dfrac{66}{35} = 1\dfrac{31}{35}$
7.	$\dfrac{5}{13} \div \dfrac{10}{13}$	$\dfrac{5}{10} = \dfrac{1}{2}$	22.	$2\dfrac{4}{9} \div 1\dfrac{3}{5}$	$\dfrac{110}{72} = \dfrac{55}{36} = 1\dfrac{19}{36}$
8.	$\dfrac{7}{8} \div \dfrac{13}{16}$	$\dfrac{14}{13} = 1\dfrac{1}{13}$	23.	$1\dfrac{2}{9} \div 3\dfrac{2}{5}$	$\dfrac{55}{153}$
9.	$\dfrac{3}{5} \div \dfrac{7}{10}$	$\dfrac{6}{7}$	24.	$2\dfrac{2}{3} \div 3$	$\dfrac{8}{9}$
10.	$\dfrac{9}{30} \div \dfrac{3}{5}$	$\dfrac{9}{18} = \dfrac{1}{2}$	25.	$1\dfrac{3}{4} \div 2\dfrac{2}{5}$	$\dfrac{35}{48}$
11.	$\dfrac{1}{3} \div \dfrac{4}{5}$	$\dfrac{5}{12}$	26.	$4 \div 1\dfrac{2}{9}$	$\dfrac{36}{11} = 3\dfrac{3}{11}$
12.	$\dfrac{2}{5} \div \dfrac{3}{4}$	$\dfrac{8}{15}$	27.	$3\dfrac{1}{5} \div 6$	$\dfrac{16}{30} = \dfrac{8}{15}$
13.	$\dfrac{3}{4} \div \dfrac{5}{9}$	$\dfrac{27}{20} = 1\dfrac{7}{20}$	28.	$2\dfrac{5}{6} \div 1\dfrac{1}{3}$	$\dfrac{51}{24} = 2\dfrac{3}{24} = 2\dfrac{1}{8}$
14.	$\dfrac{4}{5} \div \dfrac{7}{12}$	$\dfrac{48}{35} = 1\dfrac{13}{35}$	29.	$10\dfrac{2}{3} \div 8$	$\dfrac{32}{24} = \dfrac{4}{3} = 1\dfrac{1}{3}$
15.	$\dfrac{3}{8} \div \dfrac{5}{2}$	$\dfrac{6}{40} = \dfrac{3}{20}$	30.	$15 \div 2\dfrac{3}{5}$	$\dfrac{75}{13} = 5\dfrac{10}{13}$

Lesson 8:   Replacing Numbers with Letters

EUREKA MATH

Number Correct: _____
Improvement: _____

Division of Fractions II—Round 2

Directions: Determine the quotient of the fractions and simplify.

1.	$\dfrac{10}{2} \div \dfrac{5}{2}$	
2.	$\dfrac{6}{5} \div \dfrac{3}{5}$	
3.	$\dfrac{10}{7} \div \dfrac{2}{7}$	
4.	$\dfrac{3}{8} \div \dfrac{5}{8}$	
5.	$\dfrac{1}{4} \div \dfrac{3}{12}$	
6.	$\dfrac{7}{5} \div \dfrac{3}{10}$	
7.	$\dfrac{8}{15} \div \dfrac{4}{5}$	
8.	$\dfrac{5}{6} \div \dfrac{5}{12}$	
9.	$\dfrac{3}{5} \div \dfrac{7}{9}$	
10.	$\dfrac{3}{10} \div \dfrac{3}{9}$	
11.	$\dfrac{3}{4} \div \dfrac{7}{9}$	
12.	$\dfrac{7}{10} \div \dfrac{3}{8}$	
13.	$4 \div \dfrac{4}{9}$	
14.	$\dfrac{5}{8} \div 7$	
15.	$9 \div \dfrac{2}{3}$	

16.	$\dfrac{5}{8} \div 1\dfrac{3}{4}$	
17.	$\dfrac{1}{4} \div 2\dfrac{2}{5}$	
18.	$2\dfrac{3}{5} \div \dfrac{3}{8}$	
19.	$1\dfrac{3}{5} \div \dfrac{2}{9}$	
20.	$4 \div 2\dfrac{3}{8}$	
21.	$1\dfrac{1}{2} \div 5$	
22.	$3\dfrac{1}{3} \div 1\dfrac{3}{4}$	
23.	$2\dfrac{2}{5} \div 1\dfrac{1}{4}$	
24.	$3\dfrac{1}{2} \div 2\dfrac{2}{3}$	
25.	$1\dfrac{4}{5} \div 2\dfrac{3}{4}$	
26.	$3\dfrac{1}{6} \div 1\dfrac{3}{5}$	
27.	$3\dfrac{3}{5} \div 2\dfrac{1}{8}$	
28.	$5 \div 1\dfrac{1}{6}$	
29.	$3\dfrac{3}{4} \div 5\dfrac{1}{2}$	
30.	$4\dfrac{2}{3} \div 5\dfrac{1}{4}$	

Division of Fractions II—Round 2 [KEY]

Directions: Determine the quotient of the fractions and simplify.

1.	$\dfrac{10}{2} \div \dfrac{5}{2}$	$\dfrac{10}{5} = 2$	16.	$\dfrac{5}{8} \div 1\dfrac{3}{4}$	$\dfrac{20}{56} = \dfrac{5}{14}$
2.	$\dfrac{6}{5} \div \dfrac{3}{5}$	$\dfrac{6}{3} = 2$	17.	$\dfrac{1}{4} \div 2\dfrac{2}{5}$	$\dfrac{5}{48}$
3.	$\dfrac{10}{7} \div \dfrac{2}{7}$	$\dfrac{10}{2} = 5$	18.	$2\dfrac{3}{5} \div \dfrac{3}{8}$	$\dfrac{104}{15} = 6\dfrac{14}{15}$
4.	$\dfrac{3}{8} \div \dfrac{5}{8}$	$\dfrac{3}{5}$	19.	$1\dfrac{3}{5} \div \dfrac{2}{9}$	$\dfrac{72}{10} = 7\dfrac{2}{10} = 7\dfrac{1}{5}$
5.	$\dfrac{1}{4} \div \dfrac{3}{12}$	$\dfrac{3}{3} = 1$	20.	$4 \div 2\dfrac{3}{8}$	$\dfrac{32}{19} = 1\dfrac{13}{19}$
6.	$\dfrac{7}{5} \div \dfrac{3}{10}$	$\dfrac{14}{3} = 4\dfrac{2}{3}$	21.	$1\dfrac{1}{2} \div 5$	$\dfrac{3}{10}$
7.	$\dfrac{8}{15} \div \dfrac{4}{5}$	$\dfrac{8}{12} = \dfrac{2}{3}$	22.	$3\dfrac{1}{3} \div 1\dfrac{3}{4}$	$\dfrac{40}{21} = 1\dfrac{19}{21}$
8.	$\dfrac{5}{6} \div \dfrac{5}{12}$	$\dfrac{10}{5} = 2$	23.	$2\dfrac{2}{5} \div 1\dfrac{1}{4}$	$\dfrac{48}{25} = 1\dfrac{23}{25}$
9.	$\dfrac{3}{5} \div \dfrac{7}{9}$	$\dfrac{27}{35}$	24.	$3\dfrac{1}{2} \div 2\dfrac{2}{3}$	$\dfrac{21}{16} = 1\dfrac{5}{16}$
10.	$\dfrac{3}{10} \div \dfrac{3}{9}$	$\dfrac{27}{30} = \dfrac{9}{10}$	25.	$1\dfrac{4}{5} \div 2\dfrac{3}{4}$	$\dfrac{36}{55}$
11.	$\dfrac{3}{4} \div \dfrac{7}{9}$	$\dfrac{27}{28}$	26.	$3\dfrac{1}{6} \div 1\dfrac{3}{5}$	$\dfrac{95}{48} = 1\dfrac{47}{48}$
12.	$\dfrac{7}{10} \div \dfrac{3}{8}$	$\dfrac{56}{30} = \dfrac{28}{15} = 1\dfrac{13}{15}$	27.	$3\dfrac{3}{5} \div 2\dfrac{1}{8}$	$\dfrac{144}{85} = 1\dfrac{59}{85}$
13.	$4 \div \dfrac{4}{9}$	$\dfrac{36}{4} = 9$	28.	$5 \div 1\dfrac{1}{6}$	$\dfrac{30}{7} = 4\dfrac{2}{7}$
14.	$\dfrac{5}{8} \div 7$	$\dfrac{5}{56}$	29.	$3\dfrac{3}{4} \div 5\dfrac{1}{2}$	$\dfrac{30}{44} = \dfrac{15}{22}$
15.	$9 \div \dfrac{2}{3}$	$\dfrac{27}{2} = 13\dfrac{1}{2}$	30.	$4\dfrac{2}{3} \div 5\dfrac{1}{4}$	$\dfrac{56}{63} = \dfrac{8}{9}$

©2015 Great Minds. eureka-math.org
G6-M4-TE-B4-1.3.1-01.2016

EUREKA MATH™

Mathematics Curriculum

6 **GRADE**

Topic D

Expanding, Factoring, and Distributing Expressions

6.EE.A.2a, 6.EE.A.2b, 6.EE.A.3, 6.EE.A.4

Focus Standards:	6.EE.A.2	Write, read, and evaluate expressions in which letters stand for numbers.
		a. Write expressions that record operations with numbers and with letters standing for numbers. *For example, express the calculation "Subtract y from 5" as $5 - y$.*
		b. Identify parts of an expression using mathematical terms (sum, term, product, factor, quotient, coefficient); view one or more parts of an expression as a single entity. *For example, describe the expression $2(8 + 7)$ as a product of two factors; view $(8 + 7)$ as both a single entity and a sum of two terms.*
	6.EE.A.3	Apply the properties of operations to generate equivalent expressions. *For example, apply the distributive property to the expression $3(2 + x)$ to produce the equivalent expression $6 + 3x$; apply the distributive property to the expression $24x + 18y$ to produce the equivalent expression $6(4x + 3y)$; apply properties of operations to $y + y + y$ to produce the equivalent expression $3y$.*
	6.EE.A.4	Identify when two expressions are equivalent (i.e., when the two expressions name the same number regardless of which value is substituted into them). *For example, the expressions $y + y + y$ and $3y$ are equivalent because they name the same number regardless of which number y stands for.*
Instructional Days:	6	
Lesson 9:	Writing Addition and Subtraction Expressions (P)[1]	
Lesson 10:	Writing and Expanding Multiplication Expressions (P)	
Lesson 11:	Factoring Expressions (P)	
Lesson 12:	Distributing Expressions (P)	
Lessons 13–14:	Writing Division Expressions (P, P)	

---

[1]Lesson Structure Key: **P**-Problem Set Lesson, **M**-Modeling Cycle Lesson, **E**-Exploration Lesson, **S**-Socratic Lesson

Topic D: Expanding, Factoring, and Distributing Expressions

97

In Topic D, students formally utilize their understanding of expressions in order to expand, factor, and distribute. In Lesson 9, students write expressions that record addition and subtraction operations with numbers through the use of models. With a bar diagram, students understand that any number a plus any number b is the same as adding the numbers $b + a$. Students also use bar diagrams to differentiate between the mathematical terms *subtract* and *subtract from*. For instance, when subtracting b from a, they know they must first represent a in order to take away b, leading to an understanding that the expression must be written $a - b$.

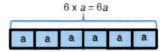

This concept deters students from writing the incorrect expression $b - a$, which is a common misconception because the number b is heard first in the expression "subtract b from a." Students continue to write expressions by combining operations with the use of parentheses.

In Lesson 10, students identify parts of an expression using mathematical terms for multiplication. They view one or more parts of an expression as a single entity. They determine that through the use of models, when a is represented 6 times, the expression is written as one entity: $6 \times a$, $6 \cdot a$, or $6a$.

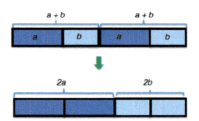

In Lesson 11, students bring with them their previous knowledge of GCF and the distributive property from Module 2 to model and write expressions using the distributive property. They move from a factored form to an expanded form of an expression, while in Lesson 12, they move from an expanded form to a factored form. In Lesson 11, students are capable of moving from the expression $2a + 2b$ to $a + b$ written twice as $(a + b) + (a + b)$ and conclude that $2a + 2b = 2(a + b)$. Conversely, students determine in Lesson 12 that $(a + b) + (a + b) = 2a + 2b$ through the following model:

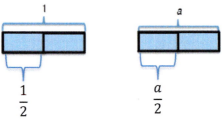

Finally, in Lessons 13 and 14, students write division expressions in two forms: dividend ÷ divisor and $\dfrac{\text{dividend}}{\text{divisor}}$, noting the relationship between the two. They determine from the model below that $1 \div 2$ is the same as $\dfrac{1}{2}$. They make an intuitive connection to expressions with letters and also determine that $a \div 2$ is the same as $\dfrac{a}{2}$.

EUREKA
MATH™

Lesson 9: Writing Addition and Subtraction Expressions

Student Outcomes

- Students write expressions that record addition and subtraction operations with numbers.

Lesson Notes

Individual white boards are recommended for this lesson.

Classwork

Example 1 (3 minutes)

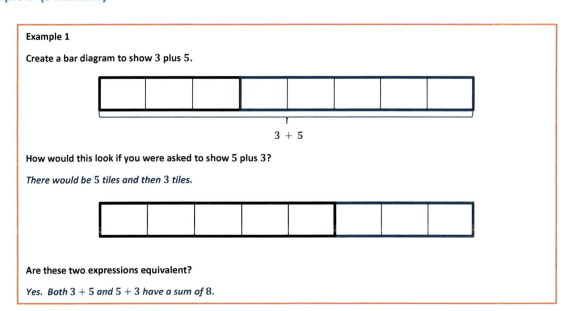

Example 1

Create a bar diagram to show 3 plus 5.

$$3 + 5$$

How would this look if you were asked to show 5 plus 3?

There would be 5 tiles and then 3 tiles.

MP.2

Are these two expressions equivalent?

Yes. Both $3 + 5$ and $5 + 3$ have a sum of 8.

Example 2 (3 minutes)

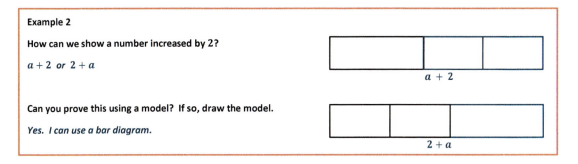

Example 2

How can we show a number increased by 2?

$a + 2$ or $2 + a$

$$a + 2$$

Can you prove this using a model? If so, draw the model.

Yes. I can use a bar diagram.

$$2 + a$$

Example 3 (3 minutes)

> **Example 3**
>
> Write an expression to show the sum of m and k.
>
> $m + k$ or $k + m$
>
>
> Which property can be used in Examples 1–3 to show that both expressions given are equivalent?
>
> *The commutative property of addition*

Example 4 (3 minutes)

> **Example 4**
>
> How can we show 10 minus 6?
>
> - Draw a bar diagram to model this expression.
>
>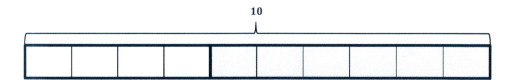
>
> - What expression would represent this model?
>
> $10 - 6$
>
> - Could we also use $6 - 10$?
>
>   *No. If we started with 6 and tried to take 10 away, the models would not match.*

MP.2

Example 5 (3 minutes)

> **Example 5**
>
> How can we write an expression to show 3 less than a number?
>
> - Start by drawing a diagram to model the subtraction. Are we taking away from the 3 or the unknown number?
>
>   *We are taking 3 away from the unknown number.*
>
>

- We are starting with some number and then subtracting 3.

> - **What expression would represent this model?**
>
>   *The expression is $n - 3$.*

Example 6 (3 minutes)

> **Example 6**
>
> **How would we write an expression to show the number c being subtracted from the sum of a and b?**
>
> - **Start by writing an expression for "the sum of a and b."**
>
> $a + b$ *or* $b + a$
>
> - **Now, show c being subtracted from the sum.**
>
> $a + b - c$ *or* $b + a - c$

Example 7 (3 minutes)

MP.2

> **Example 7**
>
> **Write an expression to show c minus the sum of a and b.**
>
> $c - (a + b)$
>
> **Why are parentheses necessary in this example and not the others?**
>
> *Without the parentheses, only a is being taken away from c, where the expression says that $a + b$ should be taken away from c.*
>
> **Replace the variables with numbers to see if $c - (a + b)$ is the same as $c - a + b$.**

If students do not see the necessity for the parentheses, have them replace the variables with numbers to see whether $c - (a + b)$ is the same as $c - a + b$.

Here is a sample of what they could try:

$$a = 1, b = 2, c = 3$$

$$
\begin{array}{ccc}
3 - (1 + 2) & & 3 - 1 + 2 \\
3 - 3 & \text{or} & 2 + 2 \\
0 & & 4
\end{array}
$$

Exercises (12 minutes)

These questions can be done on the worksheet. However, if white boards, small chalkboards, or some other personal board is available, the teacher can give instant feedback as students show their boards after each question.

Exercises

1. Write an expression to show the sum of 7 and 1.5.

    $7 + 1.5 \text{ or } 1.5 + 7$

2. Write two expressions to show w increased by 4. Then, draw models to prove that both expressions represent the same thing.

    $w + 4 \text{ and } 4 + w$

3. Write an expression to show the sum of a, b, and c.

    Answers will vary. Below are possible answers.

    $a + b + c \qquad b + c + a \qquad c + b + a$

    $a + c + b \qquad b + a + c \qquad c + a + b$

4. Write an expression and a model showing 3 less than p.

    $p - 3$

    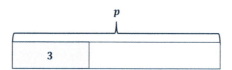

5. Write an expression to show the difference of 3 and p.

    $3 - p$

6. Write an expression to show 4 less than the sum of g and 5.

    $g + 5 - 4 \text{ or } 5 + g - 4$

7. Write an expression to show 4 decreased by the sum of g and 5.

    $4 - (g + 5) \text{ or } 4 - (5 + g)$

8. Should Exercises 6 and 7 have different expressions? Why or why not?

    The expressions are different because one includes the word "decreased by," and the other has the words "less than." The words "less than" give the amount that was taken away first, whereas the word "decreased by" gives us a starting amount and then the amount that was taken away.

©2015 Great Minds. eureka-math.org
G6-M4-TE-B4-1.3.1-01.2016

EUREKA
MATH

Closing (7 minutes)

- Write the following in words.

    □ $m + k$ *Answers will vary; the sum of m and k*

    □ $k + m$ *Answers will vary; the sum of k and m*

    □ $m - k$ *Answers will vary; m minus k*

    □ $k - m$ *Answers will vary; k minus m*

- Is $m + k$ equivalent to $k + m$? Is $m - k$ equivalent to $k - m$? Explain.

    □   *$m + k$ is equivalent to $k + m$. Both of these expressions would evaluate to the same number regardless of the numbers substituted in for k and m. However, $m - k$ and $k - m$ will not have the same result. I would be starting with a new total amount and taking away a different amount as well. The values of each expression would be different, so the expressions would not be equivalent. For example, $4 + 6 = 10$, and $6 + 4 = 10$. However, $6 - 4 = 2$, but $4 - 6 \neq 2$.*

Exit Ticket (5 minutes)

©2015 Great Minds. eureka-math.org
G6-M4-TE-B4-1.3.1-01.2016

Name _____   Date _____

Lesson 9: Writing Addition and Subtraction Expressions

Exit Ticket

1. Write an expression showing the sum of 8 and a number f.

2. Write an expression showing 5 less than the number k.

3. Write an expression showing the sum of a number h and a number w minus 11.

©2015 Great Minds. eureka-math.org
G6-M4-TE-B4-1.3.1-01.2016

Exit Ticket Sample Solutions

1. Write an expression showing the sum of 8 and a number f.

   $8 + f$ or $f + 8$

2. Write an expression showing 5 less than the number k.

   $k - 5$

3. Write an expression showing the sum of a number h and a number w minus 11.

   $h + w - 11$

Problem Set Sample Solutions

1. Write two expressions to show a number increased by 11. Then, draw models to prove that both expressions represent the same thing.

   $a + 11$ and $11 + a$

a	11

11	a

2. Write an expression to show the sum of x and y.

   $x + y$ or $y + x$

3. Write an expression to show h decreased by 13.

   $h - 13$

4. Write an expression to show k less than 3.5.

   $3.5 - k$

5. Write an expression to show the sum of g and h reduced by 11.

   $g + h - 11$

6. Write an expression to show 5 less than y, plus g.

   $y - 5 + g$

7. Write an expression to show 5 less than the sum of y and g.

   $y + g - 5$

Lesson 10: Writing and Expanding Multiplication Expressions

Student Outcomes

- Students identify parts of an expression using mathematical terms for multiplication. They view one or more parts of an expression as a single entity.

Classwork

Discussion (4 minutes)

- When we want to show multiplication of two numbers, like 5 and 7, we typically write 5×7, using the "×" to show the operation. When we start to use variables with multiplication, we can use other forms.

    $a \times b$

    $a \cdot b$

    ab

    $(a)(b)$

- Why might we want to use a form other than the × when variables are involved?
    - *The × can be confused for a variable instead of a symbol for an operation.*

- Which of the three models can be used to show multiplication where there are no variables involved?
    - *$5 \times 7, 5 \cdot 7,$ and $(5)(7)$, but not 57 because it looks like the number fifty-seven and not five times seven.*

Example 1 (10 minutes)

- When writing expressions using the fewest number of symbols, we have to refrain from using the symbols ×, ·, or ( ).

- We will also be using math terms to describe expressions and the parts of an expression. We will be using words like *factor, product, quotient, coefficient,* and *term.*

 MP.7
- A term is a part of an expression that can be added to or subtracted from the rest of the expression. In the expression $7g + 8h + 3$, what are examples of terms?
    - *$7g$, $8h$, and 3 are all terms.*

- A coefficient is a constant factor in a variable term. For example, in the term $4m$, 4 is the coefficient, and it is multiplied with m.

 EUREKA MATH

Example 1

Write each expression using the fewest number of symbols and characters. Use math terms to describe the expressions and parts of the expressions.

    a. $6 \times b$

          $6b$; the 6 is the coefficient and a factor, and the b is the variable and a factor. We can call $6b$ the product, and we can also call it a term.

    b. $4 \cdot 3 \cdot h$

          $12h$; the 12 is the coefficient and a factor, and the h is the variable and a factor. We can call $12h$ the product, and we can also call it a term.

    c. $2 \times 2 \times 2 \times a \times b$

          $8ab$; 8 is the coefficient and a factor, a and b are both variables and factors, and $8ab$ is the product and also a term.

- Variables always follow the numbers and should be written in alphabetical order. Apply this knowledge to the examples below.

---

    d. $5 \times m \times 3 \times p$

          $15mp$; 15 is the coefficient and factor, m and p are the variables and factors, $15mp$ is the product and also a term.

MP.7

- If it is helpful, you can gather the numbers together and the variables together. You can do this because of the commutative property of multiplication.

    □ $5 \times 3 \times m \times p$

---

    e. $1 \times g \times w$

          $1gw$ or gw; g and w are the variables and factors, 1 is the coefficient and factor if it is included, and gw is the product and also a term.

- What happens when you multiply by 1?

    □   *Multiplying by 1 is an example of the identity property. Any number times 1 is equal to that number. Therefore, we don't always need to write the one because $1 \times gw = gw$.*

Example 2 (5 minutes)

Example 2

To expand multiplication expressions, we will rewrite the expressions by including the "$\cdot$" back into the expressions.

    a. $5g$

          $5 \cdot g$

Lesson 10:   Writing and Expanding Multiplication Expressions   **107**

b. $7abc$

$7 \cdot a \cdot b \cdot c$

c. $12g$

$12 \cdot g$ or $2 \cdot 2 \cdot 3 \cdot g$

d. $3h \cdot 8$

$3 \cdot h \cdot 8$

e. $7g \cdot 9h$

$7 \cdot g \cdot 9 \cdot h$ or $7 \cdot g \cdot 3 \cdot 3 \cdot h$

Example 3 (5 minutes)

Example 3

a.   **Find the product of** $4f \cdot 7g$.

- It may be easier to see how we will use the fewest number of symbols and characters by expanding the expression first.

$4 \cdot f \cdot 7 \cdot g$

- Now we can multiply the numbers and then multiply the variables.

$4 \cdot 7 \cdot f \cdot g$

$28fg$

b.   **Multiply** $3de \cdot 9yz$.

- Let's start again by expanding the expression. Then, we can rewrite the expression by multiplying the numbers and then multiplying the variables.

$3 \cdot d \cdot e \cdot 9 \cdot y \cdot z$

$3 \cdot 9 \cdot d \cdot e \cdot y \cdot z$

$27deyz$

c.   **Double the product of** $6y$ **and** $3bc$.

- We can start by finding the product of $6y$ and $3bc$.

©2015 Great Minds. eureka-math.org
G6-M4-TE-B4-1.3.1-01.2016

$6 \cdot y \cdot 3 \cdot b \cdot c$

$6 \cdot 3 \cdot b \cdot c \cdot y$

$18bcy$

MP.7   ▪   What does it mean to double something?

▫   *It means to multiply by* 2.

$2 \cdot 18bcy$

$36bcy$

Exercises (14 minutes)

Students match expressions on a BINGO board. Some of the expressions are simplified, and some are expanded. To save time, provide students with a BINGO board with some of the squares already filled in. Have the remaining answers written on a smart board, a chalkboard, or an overhead projector so that students can randomly place them on the BINGO board. If there is not enough time for the BINGO game, these questions can be used on white boards, chalkboards, or some form of personal boards.

Here are the clues to be given during the game, followed by the answers that are on the board.

Questions/Clues	Answers
1. $10m$	$2 \cdot 5 \cdot \mathrm{m}$
2. $8 \cdot 3 \cdot m$	$24m$
3. Has a coefficient of 11	$11mp$
4. $14mp$	$2 \cdot 7 \cdot m \cdot p$
5. $(3m)(9p)$	$27mp$
6. $11m \cdot 2p$	$22mp$
7. $36m$	$2 \cdot 2 \cdot 3 \cdot 3 \cdot m$
8. $2 \cdot 2 \cdot 2 \cdot 5 \cdot p$	$40p$
9. $7mp \cdot 5t$	$35mpt$
10. $18pt$	$2 \cdot 3 \cdot 3 \cdot p \cdot t$
11. $7 \cdot 2 \cdot t \cdot 2 \cdot p$	$28pt$
12. Has a coefficient of 5	$5mpt$
13. $3 \cdot 3 \cdot 5 \cdot m \cdot p$	$45mp$
14. $5m \cdot 9pt$	$45mpt$
15. $10mp \cdot 4t$	$40mpt$
16. $1mpt$	mpt
17. $45mp$	$3 \cdot 3 \cdot 5 \cdot m \cdot p$
18. $(4mp)(11)$	$44mp$
19. $54mpt$	$3 \cdot 3 \cdot 3 \cdot 2 \cdot m \cdot p \cdot t$
20. Has a coefficient of 3	$3m$

These answers have already been included on premade BINGO boards to save time. The other answers can be randomly placed in the remaining spaces.

Lesson 10:   Writing and Expanding Multiplication Expressions

21. $2 \cdot 2 \cdot 2 \cdot 3 \cdot m \cdot p$	$24mp$
22. $(5m)(3p)(2t)$	$30mpt$
23. $13mp$	$(1mp)(13)$
24. Has a coefficient of 2	$2p$

Closing (3 minutes)

- What is the difference between standard form and expanded form?

  - *When we write an expression in standard form, we get rid of the operation symbol or symbols for multiplication, and we write the factors next to each other. Sometimes we might have to multiply numbers together before writing it next to the variable or variables. When we write an expression in expanded form, we write the expression as a product of the factors using the " · " symbol for multiplication.*

- How would you describe the following terms?

  1. Factor

     - *A number or variable that is multiplied to get a product*

  2. Variable

     - *A letter used to represent a number*

  3. Product

     - *The solution when two factors are multiplied*

  4. Coefficient

     - *The numerical factor that multiplies the variable*

Lesson Summary

AN EXPRESSION IN EXPANDED FORM: An expression that is written as sums (and/or differences) of products whose factors are numbers, variables, or variables raised to whole number powers is said to be in *expanded form*. A single number, variable, or a single product of numbers and/or variables is also considered to be in expanded form.

Note: Each summand of an expression in expanded form is called a *term*, and the number found by multiplying just the numbers in a term together is called the *coefficient of the term*. After the word *term* is defined, students can be shown what it means to "collect like terms" using the distributive property.

Expressions in expanded form are analogous to polynomial expressions that are written as a sum of monomials. There are two reasons for introducing this term instead of the word *polynomial*: (1) In the Common Core State Standards, the word *polynomial* cannot be formally defined before high school, but the idea behind the word is needed much sooner. (2) The progressions are very clear about not asking problems that state, "Simplify." However, they do describe *standard form* in the progressions, so students may be asked to put their answers in standard form. To get to standard form, students are asked to expand the expression and then collect like terms.

©2015 Great Minds. eureka-math.org
G6-M4-TE-B4-1.3.1-01.2016

AN EXPRESSION IN STANDARD FORM: An expression that is in expanded form where all like terms have been collected is said to be in *standard form*.

Note: Students cannot be asked to "simplify," but they can be asked to "put an expression in standard form" or "expand the expression and collect all like terms."

Exit Ticket (4 minutes)

Name _____   Date _____

Lesson 10: Writing and Expanding Multiplication Expressions

Exit Ticket

1. Rewrite the expression in standard form (use the fewest number of symbols and characters possible).

   a. $5g \cdot 7h$

   b. $3 \cdot 4 \cdot 5 \cdot m \cdot n$

2. Name the parts of the expression. Then, write it in expanded form.

   a. $14b$

   b. $30jk$

EUREKA
MATH

Exit Ticket Sample Solutions

1. Rewrite the expression in standard form (use the fewest number of symbols and characters possible).

   a. $5g \cdot 7h$

      $35gh$

   b. $3 \cdot 4 \cdot 5 \cdot m \cdot n$

      $60mn$

2. Name the parts of the expression. Then, write it in expanded form.

   a. $14b$

      $14 \cdot b$ or $2 \cdot 7 \cdot b$

      14 is the coefficient, b is the variable, and 14b is a term and the product of $14 \times b$.

   b. $30jk$

      $30 \cdot j \cdot k$ or $2 \cdot 3 \cdot 5 \cdot j \cdot k$

      30 is the coefficient, j and k are the variables, and 30jk is a term and the product of $30 \cdot j \cdot k$.

Problem Set Sample Solutions

1. Rewrite the expression in standard form (use the fewest number of symbols and characters possible).

   a. $5 \cdot y$

      $5y$

   b. $7 \cdot d \cdot e$

      $7de$

   c. $5 \cdot 2 \cdot 2 \cdot y \cdot z$

      $20yz$

   d. $3 \cdot 3 \cdot 2 \cdot 5 \cdot d$

      $90d$

2. Write the following expressions in expanded form.

   a. $3g$

      $3 \cdot g$

   b. $11mp$

      $11 \cdot m \cdot p$

c. $20yz$

$20 \cdot y \cdot z$ *or* $2 \cdot 2 \cdot 5 \cdot y \cdot z$

d. $15abc$

$15 \cdot a \cdot b \cdot c$ *or* $3 \cdot 5 \cdot a \cdot b \cdot c$

3.   Find the product.

a. $5d \cdot 7g$

$35dg$

b. $12ab \cdot 3cd$

$36abcd$

©2015 Great Minds. eureka-math.org
G6-M4-TE-B4-1.3.1-01.2016

$2 \bullet 5 \bullet m$				$35mpt$
$45mp$	$40p$		$24m$	$2 \bullet 3 \bullet 3 \bullet p \bullet t$
	$2 \bullet 7 \bullet m \bullet p$	★		$11mp$
$28pt$			$22mp$	$2 \bullet 2 \bullet 3 \bullet 3 \bullet m$
$27mp$		$5mpt$		$45mpt$

EUREKA MATH™

$22mp$		$40p$		
	$28pt$		$2 \bullet 5 \bullet m$	$2 \bullet 2 \bullet 3 \bullet 3 \bullet m$
	$45mp$	★		$35mpt$
$24m$			$45mpt$	$27mp$
$2 \bullet 7 \bullet m \bullet p$	$5mpt$		$11mp$	$2 \bullet 3 \bullet 3 \bullet p \bullet t$

©2015 Great Minds. eureka-math.org
G6-M4-TE-B4-1.3.1-01.2016

$45mp$	$40p$			$24m$
$2 \bullet 3 \bullet 3 \bullet p \bullet t$	$5mpt$	$22mp$		
$11mp$		★	$45mpt$	$2 \bullet 2 \bullet 3 \bullet 3 \bullet m$
	$27mp$	$2 \bullet 7 \bullet m \bullet p$		$28pt$
	$2 \bullet 5 \bullet m$		$35mpt$	

EUREKA MATH™

Lesson 11: Factoring Expressions

Student Outcomes

- Students model and write equivalent expressions using the distributive property. They move from expanded form to factored form of an expression.

Classwork

Fluency Exercise (5 minutes): GCF

Sprint: Refer to the Sprints and Sprint Delivery Script sections in the Module Overview for directions on how to administer a Sprint.

Example 1 (8 minutes)

MP.7

Example 1

a. **Use the model to answer the following questions.**

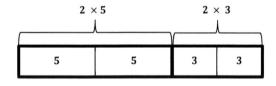

How many fives are in the model?

2

How many threes are in the model?

2

What does the expression represent in words?

The sum of two groups of five and two groups of three

What expression could we write to represent the model?

$2 \times 5 + 2 \times 3$

> *Scaffolding:*
>
> For students struggling with variables, the concept can be further solidified by having them replace the variables with whole numbers to prove that the expressions are equivalent.

b. Use the new model and the previous model to answer the next set of questions.

How many fives are in the model?

2

How many threes are in the model?

2

What does the expression represent in words?

Two groups of the sum of five and three

What expression could we write to represent the model?

$(5 + 3) + (5 + 3)$ *or* $2(5 + 3)$

c. Is the model in part (a) equivalent to the model in part (b)?

Yes, because both expressions have two 5's and two 3's. Therefore, $2 \times 5 + 2 \times 3 = 2(5 + 3)$.

d. What relationship do we see happening on either side of the equal sign?

On the left-hand side, 2 is being multiplied by 5 and then by 3 before adding the products together. On the right-hand side, the 5 and 3 are added first and then multiplied by 2.

e. In Grade 5 and in Module 2 of this year, you have used similar reasoning to solve problems. What is the name of the property that is used to say that $2(5 + 3)$ is the same as $2 \times 5 + 2 \times 3$?

The name of the property is the distributive property.

Example 2 (5 minutes)

Example 2

Now we will take a look at an example with variables. Discuss the questions with your partner.

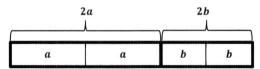

What does the model represent in words?

a plus a plus b plus b, two a's plus two b's, two times a plus two times b

EUREKA
MATH™

What does $2a$ mean?

$2a$ means that there are 2 a's or $2 \times a$.

How many a's are in the model?

2

How many b's are in the model?

2

What expression could we write to represent the model?

$2a + 2b$

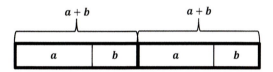

How many a's are in the expression?

2

MP.7

How many b's are in the expression?

2

What expression could we write to represent the model?

$(a + b) + (a + b) = 2(a + b)$

Are the two expressions equivalent?

Yes. Both models include 2 a's and 2 b's. Therefore, $2a + 2b = 2(a + b)$.

Example 3 (8 minutes)

Example 3

Use GCF and the distributive property to write equivalent expressions.

1. $3f + 3g =$ _____$3(f + g)$_____

What is the question asking us to do?

We need to rewrite the expression as an equivalent expression in factored form, which means the expression is written as the product of factors. The number outside of the parentheses is the GCF.

> **How would Problem 1 look if we expanded each term?**
>
> $3 \cdot f + 3 \cdot g$
>
> **What is the GCF in Problem 1?**
>
> 3
>
> **How can we use the GCF to rewrite this expression?**
>
> 3 *goes on the outside, and* $f + g$ *will go inside the parentheses.* $3(f + g)$

▪ Let's use the same ideas for Problem 2. Start by expanding the expression and naming the GCF.

> 2. $6x + 9y = \underline{\quad 3(2x + 3y) \quad}$
>
> **What is the question asking us to do?**
>
> *We need to rewrite the expression as an equivalent expression in factored form, which means the expression is written as the product of factors. The number outside of the parentheses is the GCF.*
>
> **How would Problem 2 look if we expanded each term?**
>
> $2 \cdot 3 \cdot x + 3 \cdot 3 \cdot y$
>
> **What is the GCF in Problem 2?**
>
> *The GCF is 3.*
>
> **How can we use the GCF to rewrite this expression?**
>
> *I will factor out the 3 from both terms and place it in front of the parentheses. I will place what is left in the terms inside the parentheses:* $3(2x + 3y)$.
>
> 3. $3c + 11c = \underline{\quad c(3 + 11) \quad}$
>
> **Is there a greatest common factor in Problem 3?**
>
> *Yes. When I expand, I can see that each term has a common factor c.*
>
> $3 \cdot c + 11 \cdot c$
>
> **Rewrite the expression using the distributive property.**
>
> $c(3 + 11)$
>
> 4. $24b + 8 = \underline{\quad 8(3b + 1) \quad}$

MP.7

Lesson 11:   Factoring Expressions

EUREKA
MATH™

Explain how you used GCF and the distributive property to rewrite the expression in Problem 4.

I first expanded each term. I know that 8 goes into 24, so I used it in the expansion.

$2 \cdot 2 \cdot 2 \cdot 3 \cdot b + 2 \cdot 2 \cdot 2$

I determined that $2 \cdot 2 \cdot 2$, or 8, is the common factor. So, on the outside of the parentheses I wrote 8, and on the inside I wrote the leftover factor, $3b + 1$. $8(3b + 1)$

MP.7

Why is there a 1 in the parentheses?

When I factor out a number, I am leaving behind the other factor that multiplies to make the original number. In this case, when I factor out an 8 from 8, I am left with a 1 because $8 \times 1 = 8$.

How is this related to the first two examples?

In the first two examples, we saw that we could rewrite the expressions by thinking about groups.

We can either think of $24b + 8$ as 8 groups of $3b$ and 8 groups of 1 or as 8 groups of the sum of $3b + 1$. This shows that $8(3b) + 8(1) = 8(3b + 1)$ is the same as $24b + 8$.

Exercises (12 minutes)

If times allows, have students practice these questions on white boards or small personal boards.

Exercises

1. Apply the distributive property to write equivalent expressions.

    a. $7x + 7y$

        $7(x + y)$

    b. $15g + 20h$

        $5(3g + 4h)$

    c. $18m + 42n$

        $6(3m + 7n)$

    d. $30a + 39b$

        $3(10a + 13b)$

    e. $11f + 15f$

        $f(11 + 15)$

    f. $18h + 13h$

        $h(18 + 13)$

    g. $55m + 11$

        $11(5m + 1)$

Lesson 11:   Factoring Expressions

h. $7 + 56y$

 $7(1 + 8y)$

2.   Evaluate each of the expressions below.

a. $6x + 21y$ and $3(2x + 7y)$ $x = 3$ and $y = 4$

 $6(3) + 21(4)$ $3(2 \cdot 3 + 7 \cdot 4)$

 $18 + 84$ $3(6 + 28)$

 102 $3(34)$

 102 102

b. $5g + 7g$ and $g(5 + 7)$ $g = 6$

 $5(6) + 7(6)$ $6(5 + 7)$

 $30 + 42$ $6(12)$

 72 72

c. $14x + 2$ and $2(7x + 1)$ $x = 10$

 $14(10) + 2$ $2(7 \cdot 10 + 1)$

 $140 + 2$ $2(70 + 1)$

 142 $2(71)$

 142 142

d.   Explain any patterns that you notice in the results to parts (a)–(c).

 Both expressions in parts (a)–(c) evaluated to the same number when the indicated value was substituted for the variable. This shows that the two expressions are equivalent for the given values.

e.   What would happen if other values were given for the variables?

 Because the two expressions in each part are equivalent, they evaluate to the same number, no matter what value is chosen for the variable.

EUREKA
MATH™

©2015 Great Minds. eureka-math.org
G6-M4-TE-B4-1.3.1-01.2016

Closing (3 minutes)

> **Closing**
>
> How can use you use your knowledge of GCF and the distributive property to write equivalent expressions?
>
> *We can use our knowledge of GCF and the distributive property to change expressions from standard form to factored form.*
>
> Find the missing value that makes the two expressions equivalent.
>
> $4x + 12y$ $\underline{\qquad\quad 4}\,(x + 3y)$
>
> $35x + 50y$ $\underline{\qquad\quad 5}\,(7x + 10y)$
>
> $18x + 9y$ $\underline{\qquad\quad 9}\,(2x + y)$
>
> $32x + 8y$ $\underline{\qquad\quad 8}\,(4x + y)$
>
> $100x + 700y$ $\underline{\qquad 100}\,(x + 7y)$
>
> Explain how you determine the missing number.
>
> *I would expand each term and determine the greatest common factor. The greatest common factor is the number that is placed on the blank line.*
>
> > **Lesson Summary**
> >
> > AN EXPRESSION IN FACTORED FORM: An expression that is a product of two or more expressions is said to be in *factored form*.

Exit Ticket (4 minutes)

Name _____   Date _____

Lesson 11: Factoring Expressions

Exit Ticket

Use greatest common factor and the distributive property to write equivalent expressions in factored form.

1. $2x + 8y$

2. $13ab + 15ab$

3. $20g + 24h$

Lesson 11:   Factoring Expressions

EUREKA
MATH™

Exit Ticket Sample Solutions

Use greatest common factor and the distributive property to write equivalent expressions in factored form.

1. $2x + 8y$

    $2(x + 4y)$

2. $13ab + 15ab$

    $ab(13 + 15)$

3. $20g + 24h$

    $4(5g + 6h)$

Problem Set Sample Solutions

1. Use models to prove that $3(a + b)$ is equivalent to $3a + 3b$.

    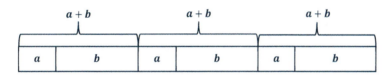

    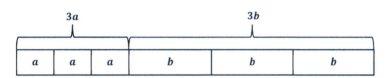

2. Use greatest common factor and the distributive property to write equivalent expressions in factored form for the following expressions.

    a. $4d + 12e$

        $4(d + 3e)$ or $4(1d + 3e)$

    b. $18x + 30y$

        $6(3x + 5y)$

    c. $21a + 28y$

        $7(3a + 4y)$

    d. $24f + 56g$

        $8(3f + 7g)$

Number Correct: _____

Greatest Common Factor—Round 1

Directions: Determine the greatest common factor of each pair of numbers.

1.	GCF of 10 and 50		16.	GCF of 45 and 72	
2.	GCF of 5 and 35		17.	GCF of 28 and 48	
3.	GCF of 3 and 12		18.	GCF of 44 and 77	
4.	GCF of 8 and 20		19.	GCF of 39 and 66	
5.	GCF of 15 and 35		20.	GCF of 64 and 88	
6.	GCF of 10 and 75		21.	GCF of 42 and 56	
7.	GCF of 9 and 30		22.	GCF of 28 and 42	
8.	GCF of 15 and 33		23.	GCF of 13 and 91	
9.	GCF of 12 and 28		24.	GCF of 16 and 84	
10.	GCF of 16 and 40		25.	GCF of 36 and 99	
11.	GCF of 24 and 32		26.	GCF of 39 and 65	
12.	GCF of 35 and 49		27.	GCF of 27 and 87	
13.	GCF of 45 and 60		28.	GCF of 28 and 70	
14.	GCF of 48 and 72		29.	GCF of 26 and 91	
15.	GCF of 50 and 42		30.	GCF of 34 and 51	

    Lesson 11:   Factoring Expressions

Greatest Common Factor—Round 1 [KEY]

Directions: Determine the greatest common factor of each pair of numbers.

1.	GCF of 10 and 50	**10**		16.	GCF of 45 and 72	**9**
2.	GCF of 5 and 35	**5**		17.	GCF of 28 and 48	**4**
3.	GCF of 3 and 12	**3**		18.	GCF of 44 and 77	**11**
4.	GCF of 8 and 20	**4**		19.	GCF of 39 and 66	**3**
5.	GCF of 15 and 35	**5**		20.	GCF of 64 and 88	**8**
6.	GCF of 10 and 75	**5**		21.	GCF of 42 and 56	**14**
7.	GCF of 9 and 30	**3**		22.	GCF of 28 and 42	**14**
8.	GCF of 15 and 33	**3**		23.	GCF of 13 and 91	**13**
9.	GCF of 12 and 28	**4**		24.	GCF of 16 and 84	**4**
10.	GCF of 16 and 40	**8**		25.	GCF of 36 and 99	**9**
11.	GCF of 24 and 32	**8**		26.	GCF of 39 and 65	**13**
12.	GCF of 35 and 49	**7**		27.	GCF of 27 and 87	**3**
13.	GCF of 45 and 60	**15**		28.	GCF of 28 and 70	**14**
14.	GCF of 48 and 72	**24**		29.	GCF of 26 and 91	**13**
15.	GCF of 50 and 42	**2**		30.	GCF of 34 and 51	**17**

Lesson 11:   Factoring Expressions

©2015 Great Minds. eureka-math.org
G6-M4-TE-B4-1.3.1-01.2016

Number Correct: _____

Improvement: _____

Greatest Common Factor—Round 2

Directions: Determine the greatest common factor of each pair of numbers.

1.	GCF of 20 and 80		16.	GCF of 33 and 99	
2.	GCF of 10 and 70		17.	GCF of 38 and 76	
3.	GCF of 9 and 36		18.	GCF of 26 and 65	
4.	GCF of 12 and 24		19.	GCF of 39 and 48	
5.	GCF of 15 and 45		20.	GCF of 72 and 88	
6.	GCF of 10 and 95		21.	GCF of 21 and 56	
7.	GCF of 9 and 45		22.	GCF of 28 and 52	
8.	GCF of 18 and 33		23.	GCF of 51 and 68	
9.	GCF of 12 and 32		24.	GCF of 48 and 84	
10.	GCF of 16 and 56		25.	GCF of 21 and 63	
11.	GCF of 40 and 72		26.	GCF of 64 and 80	
12.	GCF of 35 and 63		27.	GCF of 36 and 90	
13.	GCF of 30 and 75		28.	GCF of 28 and 98	
14.	GCF of 42 and 72		29.	GCF of 39 and 91	
15.	GCF of 30 and 28		30.	GCF of 38 and 95	

EUREKA
MATH™

Greatest Common Factor—Round 2 [KEY]

Directions: Determine the greatest common factor of each pair of numbers.

1.	GCF of 20 and 80	**20**	16.	GCF of 33 and 99	**33**
2.	GCF of 10 and 70	**10**	17.	GCF of 38 and 76	**38**
3.	GCF of 9 and 36	**9**	18.	GCF of 26 and 65	**13**
4.	GCF of 12 and 24	**12**	19.	GCF of 39 and 48	**3**
5.	GCF of 15 and 45	**15**	20.	GCF of 72 and 88	**8**
6.	GCF of 10 and 95	**5**	21.	GCF of 21 and 56	**7**
7.	GCF of 9 and 45	**9**	22.	GCF of 28 and 52	**4**
8.	GCF of 18 and 33	**3**	23.	GCF of 51 and 68	**17**
9.	GCF of 12 and 32	**4**	24.	GCF of 48 and 84	**12**
10.	GCF of 16 and 56	**8**	25.	GCF of 21 and 63	**21**
11.	GCF of 40 and 72	**8**	26.	GCF of 64 and 80	**16**
12.	GCF of 35 and 63	**7**	27.	GCF of 36 and 90	**18**
13.	GCF of 30 and 75	**15**	28.	GCF of 28 and 98	**14**
14.	GCF of 42 and 72	**6**	29.	GCF of 39 and 91	**13**
15.	GCF of 30 and 28	**2**	30.	GCF of 38 and 95	**19**

Lesson 11:   Factoring Expressions

Lesson 12: Distributing Expressions

Student Outcomes

▪ Students model and write equivalent expressions using the distributive property. They move from the factored form to the expanded form of an expression.

Classwork

Opening Exercise (3 minutes)

Opening Exercise

a. **Create a model to show** 2×5.

5	5

b. **Create a model to show** $2 \times b$, **or** $2b$.

b	b

Example 1 (8 minutes)

Example 1

Write an expression that is equivalent to $2(a + b)$.

▪ In this example, we have been given the factored form of the expression.
▪ To answer this question, we can create a model to represent $2(a + b)$.
▪ Let's start by creating a model to represent $(a + b)$.

MP.7

Create a model to represent $(a + b)$.

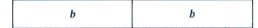

The expression $2(a + b)$ **tells us that we have 2 of the** $(a + b)$**'s. Create a model that shows 2 groups of** $(a + b)$.

EUREKA MATH™

How many a's and how many b's do you see in the diagram?

There are 2 a's and 2 b's.

How would the model look if we grouped together the a's and then grouped together the b's?

What expression could we write to represent the new diagram?

$2a + 2b$

- This expression is written in expanded form.

What conclusion can we draw from the models about equivalent expressions?

$2(a + b) = 2a + 2b$

- To prove that these two forms are equivalent, let's substitute some values for a and b and see what happens.

MP.7

Let $a = 3$ and $b = 4$.

$2(a + b)$	$2a + 2b$
$2(3 + 4)$	$2(3) + 2(4)$
$2(7)$	$6 + 8$
14	14

- Note: If students do not believe yet that the two are equal, continue substituting more values for a and b until they are convinced.

What happens when we double $(a + b)$?

We double a, and we double b.

Example 2 (5 minutes)

Example 2

Write an expression that is equivalent to double $(3x + 4y)$.

How can we rewrite double $(3x + 4y)$?

Double is the same as multiplying by two.

$2(3x + 4y)$ or $6x + 8y$

Is this expression in factored form, expanded form, or neither?

The first expression is in factored form, and the second expression is in expanded form.

Let's start this problem the same way that we started the first example. What should we do?

We can make a model of $3x + 4y$.

3x	4y

How can we change the model to show $2(3x + 4y)$?

We can make two copies of the model.

Are there terms that we can combine in this example?

MP.7

Yes. There are 6 x's and 8 y's.

So, the model is showing $6x + 8y$.

What is an equivalent expression that we can use to represent $2(3x + 4y)$?

$2(3x + 4y) = 6x + 8y$

This is the same as $2(3x) + 2(4y)$.

Summarize how you would solve this question without the model.

When there is a number outside the parentheses, I would multiply it by all the terms on the inside of the parentheses.

Example 3 (3 minutes)

Example 3

Write an expression in expanded form that is equivalent to the model below.

What factored expression is represented in the model?

$y(4x + 5)$

EUREKA
MATH

How can we rewrite this expression in expanded form?

$y(4x) + y(5)$

$4xy + 5y$

Example 4 (3 minutes)

MP.7

- How can we use our work in the previous examples to write the following expression?

Example 4

Write an expression in expanded form that is equivalent to $3(7d + 4e)$.

We will multiply $3 \times 7d$ and $3 \times 4e$.

We would get $21d + 12e$. So, $3(7d + 4e) = 21d + 12e$.

Exercises (15 minutes)

Exercises

Create a model for each expression below. Then, write another equivalent expression using the distributive property.

1. $3(x + y)$

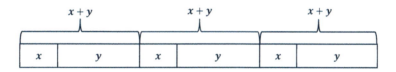

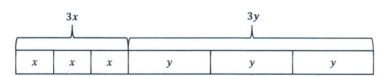

$3x + 3y$

2. $4(2h + g)$

$8h + 4g$

Apply the distributive property to write equivalent expressions in expanded form.

3. $8(h + 3)$

    $8h + 24$

4. $3(2h + 7)$

    $6h + 21$

5. $5(3x + 9y)$

    $15x + 45y$

6. $4(11h + 3g)$

    $44h + 12g$

7.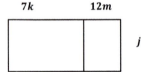

    $7jk + 12jm$

8. $a(9b + 13)$

    $9ab + 13a$

Closing (3 minutes)

- State what the expression $a(b + c)$ represents.
    - *a groups of the quantity b plus c*
- Explain in your own words how to write an equivalent expression in expanded form when given an expression in the form of $a(b + c)$. Then, create your own example to show off what you know.
    - *To write an equivalent expression, I would multiply a times b and a times c. Then, I would add the two products together.*
    - *Examples will vary.*
- State what the equivalent expression in expanded form represents.
    - *$ab + ac$ means a groups of size b plus a groups of size c.*

Exit Ticket (5 minutes)

EUREKA
MATH™

Name _____   Date _____

Lesson 12: Distributing Expressions

Exit Ticket

Use the distributive property to write the following expressions in expanded form.

1. $2(b + c)$

2. $5(7h + 3m)$

3. $e(f + g)$

Exit Ticket Sample Solutions

Use the distributive property to write the following expressions in expanded form.

1. $2(b + c)$

    $2b + 2c$

2. $5(7h + 3m)$

    $35h + 15m$

3. $e(f + g)$

    $ef + eg$

Problem Set Sample Solutions

1. Use the distributive property to write the following expressions in expanded form.

    a. $4(x + y)$

        $4x + 4y$

    b. $8(a + 3b)$

        $8a + 24b$

    c. $3(2x + 11y)$

        $6x + 33y$

    d. $9(7a + 6b)$

        $63a + 54b$

    e. $c(3a + b)$

        $3ac + bc$

    f. $y(2x + 11z)$

        $2xy + 11yz$

EUREKA
MATH™

2.   Create a model to show that $2(2x + 3y) = 4x + 6y$.

Lesson 13: Writing Division Expressions

Student Outcomes

- Students write numerical expressions in two forms, "dividend ÷ divisor" and $\frac{\text{“dividend”}}{\text{divisor}}$, and note the relationship between the two.

Lesson Notes

This is day one of a two-day lesson.

Classwork

Discussion (8 minutes)

The discussion serves as a chance for students to show what they know about division and what division looks like. The discussion should conclude with the overall idea that writing $a \div b$ as $\frac{a}{b}$ is a strategic format when working algebraically.

- How can we write or show 8 divided by 2? (Students may be allowed to explain or even draw examples for the class to see.)
  - *Answers will vary. Students can draw models, arrays, use the division symbol, and some may even use a fraction.*
- When working with algebraic expressions, are any of these expressions or models more efficient than others?
  - *Writing a fraction to show division is more efficient.*
- Is $\frac{8}{2}$ the same as $\frac{2}{8}$?
  - *No, they are not the same. $\frac{8}{2} = 4$, while $\frac{2}{8} = \frac{1}{4}$.*
- How would we show a divided by b using a fraction?
  - $\frac{a}{b}$

MP.6

Example 1 (5 minutes)

Example 1

Write an expression showing $1 \div 2$ without the use of the division symbol.

- Let's start by looking at a model of $1 \div 2$.
  - □   *We can make a bar diagram.*

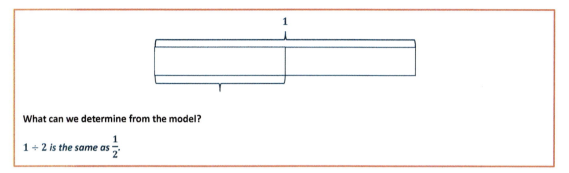

What can we determine from the model?

$1 \div 2$ *is the same as* $\dfrac{1}{2}$.

Example 2 (5 minutes)

Example 2

Write an expression showing $a \div 2$ without the use of the division symbol.

- Here we have a variable being divided by 2. Let's start by looking at a model of $a \div 2$.
  - □   *We can make a bar diagram.*

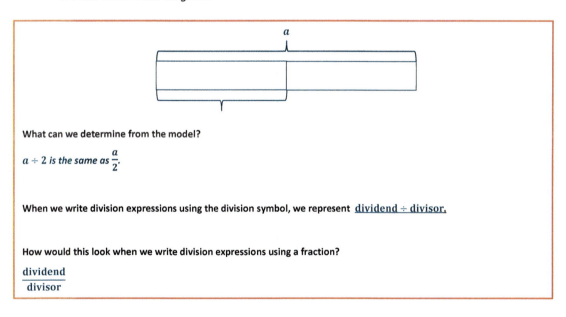

What can we determine from the model?

$a \div 2$ *is the same as* $\dfrac{a}{2}$.

When we write division expressions using the division symbol, we represent dividend ÷ divisor.

How would this look when we write division expressions using a fraction?

$\dfrac{\text{dividend}}{\text{divisor}}$

Example 3 (8 minutes)

Example 3

   a.   Write an expression showing $a \div b$ without the use of the division symbol.

- How can we use what we just learned in Examples 1 and 2 to help us with this example?
  - *The dividend is the numerator, and the divisor is the denominator.*

$$\frac{a}{b}$$

  b.   **Write an expression for g divided by the quantity h plus 3.**

- How would this look with the division symbol?
  - $g \div (h + 3)$
- Now, let's rewrite this using a fraction.

MP.6

$$\frac{g}{h + 3}$$

  c.   **Write an expression for the quotient of the quantity m reduced by 3 and 5.**

- Let's start again by writing this using a division symbol first.
  - $(m - 3) \div 5$
- Next, we will rewrite it using the fraction bar.

$$\frac{m - 3}{5}$$

Exercises (10 minutes)

Have students use a white board or small board to practice the following questions.

Exercises

Write each expression two ways: using the division symbol and as a fraction.

  a.   12 divided by 4

  $12 \div 4$ *and* $\dfrac{12}{4}$

  b.   3 divided by 5

  $3 \div 5$ *and* $\dfrac{3}{5}$

  c. a divided by 4

  $a \div 4$ *and* $\dfrac{a}{4}$

EUREKA
MATH™

d. The quotient of 6 and m

$6 \div m$ and $\dfrac{6}{m}$

e. Seven divided by the quantity x plus y

$7 \div (x + y)$ and $\dfrac{7}{x + y}$

f. y divided by the quantity x minus 11

$y \div (x - 11)$ and $\dfrac{y}{x - 11}$

g. The sum of the quantity h and 3 divided by 4

$(h + 3) \div 4$ and $\dfrac{h + 3}{4}$

h. The quotient of the quantity k minus 10 and m

$(k - 10) \div m$ and $\dfrac{k - 10}{m}$

Closing (4 minutes)

▪ Explain to your neighbor how you would rewrite any division problem using a fraction.

□ *The dividend would become the numerator, and the divisor would become the denominator.*

Exit Ticket (5 minutes)

Lesson 13:   Writing Division Expressions

Name _____   Date _____

Lesson 13: Writing Division Expressions

Exit Ticket

Rewrite the expressions using the division symbol and as a fraction.

1.   The quotient of m and 7

2.   Five divided by the sum of a and b

3.   The quotient of k decreased by 4 and 9

Exit Ticket Sample Solutions

Rewrite the expressions using the division symbol and as a fraction.

1. The quotient of m and 7

    $m \div 7$ and $\dfrac{m}{7}$

2. Five divided by the sum of a and b

    $5 \div (a + b)$ and $\dfrac{5}{a + b}$

3. The quotient of k decreased by 4 and 9

    $(k - 4) \div 9$ and $\dfrac{k - 4}{9}$

Problem Set Sample Solutions

1. Rewrite the expressions using the division symbol and as a fraction.

    a. Three divided by 4

        $3 \div 4$ and $\dfrac{3}{4}$

    b. The quotient of m and 11

        $m \div 11$ and $\dfrac{m}{11}$

    c. 4 divided by the sum of h and 7

        $4 \div (h + 7)$ and $\dfrac{4}{h + 7}$

    d. The quantity x minus 3 divided by y

        $(x - 3) \div y$ and $\dfrac{x - 3}{y}$

2. Draw a model to show that $x \div 3$ is the same as $\dfrac{x}{3}$.

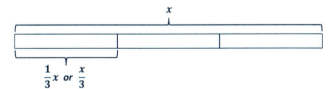

$\dfrac{1}{3}x$ or $\dfrac{x}{3}$

Lesson 14: Writing Division Expressions

Student Outcomes

- Students write numerical expressions in two forms, "dividend ÷ divisor" and $\dfrac{\text{"dividend"}}{\text{divisor}}$, and note the relationship between the two.

Lesson Notes

This is the second day of a two-day lesson.

Classwork

Fluency Exercise (5 Minutes): Long Division Algorithm

RWBE: Refer to the Rapid White Board Exchanges sections in the Module Overview for directions on how to administer an RWBE.

Example 1 (5 minutes)

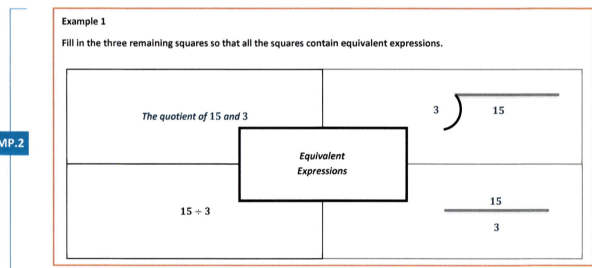

Example 1

Fill in the three remaining squares so that all the squares contain equivalent expressions.

The quotient of 15 *and* 3

Equivalent Expressions

$3\,\overline{)\,15}$

$15 \div 3$

$\dfrac{15}{3}$

EUREKA MATH™

Example 2 (5 minutes)

MP.2

Example 2

Fill in a blank copy of the four boxes using the words *dividend* and *divisor* so that it is set up for any example.

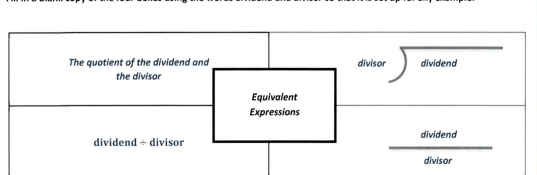

Exercises (20 minutes)

Students work in pairs. Each pair is given a set of expressions to work on. There are several different versions that can be printed and used so that a variety of questions can be used throughout the classroom. Students fill in the four rectangles, one with the given information and three with equivalent expressions.

Exercises

Complete the missing spaces in each rectangle set.

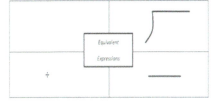 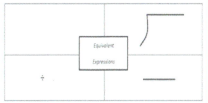

> *Scaffolding:*
>
> The sets of eight questions used in the Exercises can be tailored to fit the level at which students are working. A set for lower-level learners and/or a set for advanced learners may be written as needed.

Set A

Answers:

1. $5 \div p$ 5 divided by p, $\dfrac{5}{p}$, $p \overline{)5}$

2.   The quotient of g and h $g \div h$, $\dfrac{g}{h}$, $h \overline{)g}$

3. $w \overline{)23}$ 23 divided by w, $23 \div w$, $\dfrac{23}{w}$

4. $\dfrac{y}{x+8}$ y divided by the sum of x and 8, $y \div (x+8)$, $x+8 \overline{)y}$

5.   7 divided by the quantity a minus 6 $7 \div (a-6)$, $\dfrac{7}{a-6}$, $a-6 \overline{)7}$

6. $3 \overline{)m+11}$ The sum of m and 11 divided by 3, $(m+11) \div 3$, $\dfrac{m+11}{3}$

©2015 Great Minds. eureka-math.org
G6-M4-TE-B4-1.3.1-01.2016

7. $(f + 2) \div g$

The sum of f and 2 divided by g, $\dfrac{f+2}{g}$, $g\overline{)f + 2}$

8. $\dfrac{c-9}{d+3}$

The quotient of c minus 9 and d plus 3, $(c - 9) \div (d + 3)$, $d + 3\overline{)c - 9}$

Set B

Answers:

1. $h \div 11$

The quotient of h and 11, $11\overline{)h}$, $\dfrac{h}{11}$

2. The quotient of m and n

$m \div n$, $n\overline{)m}$, $\dfrac{m}{n}$

3. $5\overline{)j}$

The quotient of j and 5, $j \div 5$, $\dfrac{j}{5}$

4. $\dfrac{h}{m-4}$

h divided by the quantity m minus 4, $h \div (m - 4)$, $m - 4\overline{)h}$

5. f divided by the quantity g minus 11

$f \div (g - 11)$, $g - 11\overline{)f}$, $\dfrac{f}{g-11}$

6. $18\overline{)a + 5}$

The sum of a and 5 divided by 18, $(a + 5) \div 18$, $\dfrac{a+5}{18}$

7. $(y - 3) \div x$

The quantity y minus 3 divided by x, $x\overline{)y - 3}$, $\dfrac{y-3}{x}$

8. $\dfrac{g+5}{h-11}$

The quantity g plus 5 divided by the quantity h minus 11,

$(g + 5) \div (h - 11)$, $h - 11\overline{)g + 5}$

Set C

Answers:

1. $6 \div k$

6 divided by k, $k\overline{)6}$, $\dfrac{6}{k}$

2. The quotient of j and k

$j \div k$, $k\overline{)j}$, $\dfrac{j}{k}$

3. $10\overline{)a}$

a divided by 10, $a \div 10$, $\dfrac{a}{10}$

4. $\dfrac{15}{f-2}$

15 divided by the quantity f minus 2, $15 \div (f - 2)$, $f - 2\overline{)15}$

5. 13 divided by the sum of h and 1

$13 \div (h + 1)$, $h + 1\overline{)13}$, $\dfrac{13}{h+1}$

6. $3\overline{)c + 18}$

The sum of c plus 18 divided by 3, $(c + 18) \div 3$, $\dfrac{c + 18}{3}$

7. $(h - 2) \div m$

The quantity h minus 2 divided by m, $m\overline{)h - 2}$, $\dfrac{h-2}{m}$

8. $\dfrac{4-m}{n+11}$

The quantity 4 minus m divided by the sum of n and 11,

$(4 - m) \div (n + 11)$, $n + 11\overline{)4 - m}$

EUREKA MATH™

Closing (7 minutes)

Two pairs of students trade pages to check each other's work. If all of the boxes are correct, students write a sentence that summarizes why the expressions are equivalent. If there are mistakes, students write sentences to explain how to correct them.

Students evaluate some of the expressions. Many answers need to be written as fractions or decimals.

$$\text{Set A: } p = 3, w = 5, a = 10$$
$$\text{Set B: } h = 4, j = 8, a = 10$$
$$\text{Set C: } k = 2, a = 10, c = 6$$

Exit Ticket (3 minutes)

EUREKA
MATH™

Lesson 14:   Writing Division Expressions

149

©2015 Great Minds. eureka-math.org
G6-M4-TE-B4-1.3.1-01.2016

Name _____   Date _____

Lesson 14: Writing Division Expressions

Exit Ticket

1. Write the division expression in words and as a fraction.

$$(g + 12) \div h$$

2. Write the following division expression using the division symbol and as a fraction: f divided by the quantity h minus 3.

EUREKA MATH

Exit Ticket Sample Solutions

1. Write the division expression in words and as a fraction.

$$(g + 12) \div h$$

The sum of g and 12 divided by h, $\dfrac{g+12}{h}$

2. Write the following division expression using the division symbol and as a fraction: f divided by the quantity h minus 3.

$f \div (h - 3)$ and $\dfrac{f}{h-3}$

Problem Set Sample Solutions

Complete the missing spaces in each rectangle set.

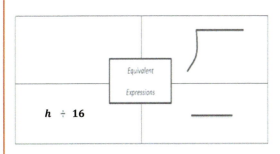

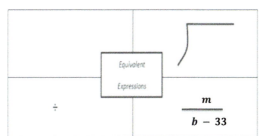

The quotient of h and 16, $16\overline{)h}$, $\dfrac{h}{16}$

m divided by the quantity b minus 33, $b - 33\overline{)m}$, $m \div (b - 33)$

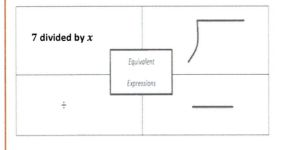

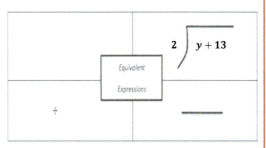

$x\overline{)7}$, $7 \div x$, $\dfrac{7}{x}$

The sum of y and 13 divided by 2, $(y + 13) \div 2$, $\dfrac{y+13}{2}$

Exercise Handout

Set A

1. $5 \div p$
2. The quotient of g and h
3. $w\overline{)23}$
4. $\dfrac{y}{x+8}$
5. 7 divided by the quantity a minus 6
6. $3\overline{)m+11}$
7. $(f + 2) \div g$
8. $\dfrac{c-9}{d+3}$

Set B

1. $h \div 11$
2. The quotient of m and n
3. $5\overline{)j}$
4. $\dfrac{h}{m-4}$
5. f divided by the quantity g minus 11
6. $18\overline{)a+5}$
7. $(y - 3) \div x$
8. $\dfrac{g+5}{h-11}$

Set C

1. $6 \div k$
2. The quotient of j and k
3. $10\overline{)a}$
4. $\dfrac{15}{f-2}$
5. 13 divided by the sum of h and 1
6. $3\overline{)c+18}$
7. $(h - 2) \div m$
8. $\dfrac{4-m}{n+11}$

Long Division Algorithm

Progression of Exercises

1. 3,282 ÷ 6

    547

2. 2,712 ÷ 3

    904

3. 15,036 ÷ 7

    2,148

4. 1,788 ÷ 8

    223.5

5. 5,736 ÷ 12

    478

6. 35,472 ÷ 16

    2,217

7. 13,384 ÷ 28

    478

8. 31,317 ÷ 39

    803

9. 1,113 ÷ 42

    26.5

10. 4,082 ÷ 52

    78.5

EUREKA
MATH™

Lesson 14:   Writing Division Expressions

153

©2015 Great Minds. eureka-math.org
G6-M4-TE-B4-1.3.1-01.2016

6
GRADE

Mathematics Curriculum

Topic E

Expressing Operations in Algebraic Form

6.EE.A.2a, 6.EE.A.2b

Focus Standards:	6.EE.A.2	Write, read, and evaluate expressions in which letters stand for numbers.
		a. Write expressions that record operations with numbers and with letters standing for numbers. *For example, express the calculation "Subtract y from 5" as $5 - y$.*
		b. Identify parts of an expression using mathematical terms (sum, term, product, factor, quotient, coefficient); view one or more parts of an expression as a single entity. *For example, describe the expression $2(8 + 7)$ as a product of two factors; view $(8 + 7)$ as both a single entity and a sum of two terms.*
Instructional Days:	3	
Lesson 15:	Read Expressions in Which Letters Stand for Numbers (P)[1]	
Lessons 16–17:	Write Expressions in Which Letters Stand for Numbers (M, P)	

In Topic E, students express mathematical terms in algebraic form. They read and write expressions in which letters stand for numbers. In Lesson 15, students provide word descriptions for operations in an algebraic expression. Given the expression $4b + c$, students assign the operation term *product* for multiplication and the term *sum* for addition. They verbalize the expression as "the sum of c and the product of 4 and b." However, in Lessons 16 and 17, students are given verbal expressions, and they write algebraic expressions to record operations with numbers and letters standing for numbers. Provided the verbal expression, "Devin quadrupled his money and deposited it with his mother's," students write the expression $4a + b$, where a represents the amount of money Devin originally had and b represents the amount of money his mother has. Or, provided the verbal expression, "Crayons and markers were put together and distributed equally to six tables," students create the algebraic expression $\frac{a+b}{6}$, where a represents the number of crayons and b represents the number of markers. Mastery of reading and writing expressions in this topic leads to a fluent transition in the next topic where students read, write, and evaluate expressions.

---

[1]Lesson Structure Key: **P**-Problem Set Lesson, **M**-Modeling Cycle Lesson, **E**-Exploration Lesson, **S**-Socratic Lesson

Lesson 15: Read Expressions in Which Letters Stand for Numbers

Student Outcomes

- Students read expressions in which letters stand for numbers. They assign operation terms to operations when reading.
- Students identify parts of an algebraic expression using mathematical terms for all operations.

Classwork

Opening Exercise (10 minutes)

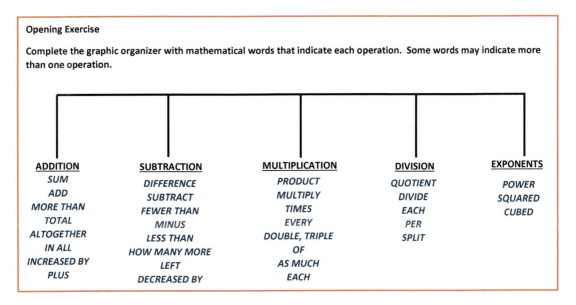

Opening Exercise

Complete the graphic organizer with mathematical words that indicate each operation. Some words may indicate more than one operation.

ADDITION	SUBTRACTION	MULTIPLICATION	DIVISION	EXPONENTS
SUM	DIFFERENCE	PRODUCT	QUOTIENT	POWER
ADD	SUBTRACT	MULTIPLY	DIVIDE	SQUARED
MORE THAN	FEWER THAN	TIMES	EACH	CUBED
TOTAL	MINUS	EVERY	PER	
ALTOGETHER	LESS THAN	DOUBLE, TRIPLE	SPLIT	
IN ALL	HOW MANY MORE	OF		
INCREASED BY	LEFT	AS MUCH		
PLUS	DECREASED BY	EACH		

Have different students share the vocabulary words they wrote in each category. If students are missing vocabulary words in their graphic organizers, have them add the new words. At the end of the Opening Exercise, every student should have the same lists of vocabulary words for each operation.

Example 1 (13 minutes)

Have students write down an expression using words. Encourage students to refer back to the graphic organizer created during the Opening Exercise. After providing students time to write each expression, have different students read each expression out loud. Each student should use different mathematical vocabulary.

Example 1

Write an expression using words.

a. $a - b$

Possible answers: a minus b; the difference of a and b; a decreased by b; b subtracted from a

b. xy

Possible answers: the product of x and y; x multiplied by y; x times y

c. $4f + p$

Possible answers: p added to the product of 4 and f; 4 times f plus p; the sum of 4 multiplied by f and p

MP.6

d. $d - b^3$

Possible answers: d minus b cubed; the difference of d and the quantity b to the third power

e. $5(u - 10) + h$

Possible answers: Add h to the product of 5 and the difference of u and 10; 5 times the quantity of u minus 10 added to h.

f. $\dfrac{3}{d+f}$

Possible answers: Find the quotient of 3 and the sum of d and f; 3 divided by the quantity d plus f.

- Why is 3 divided by d plus f not a correct answer?
  - *Possible answer: 3 divided by d plus f would indicate that we divide 3 and d first and then add f, but this is not what the expression is showing.*

Exercises (12 minutes)

Students work with a partner to complete the following problems.

Scaffolding:

If students are using the vocabulary words well or finish early, ask students to write two different expressions for Exercises 1–4.

Exercises

Circle all the vocabulary words that could be used to describe the given expression.

1. $6h - 10$

ADDITION   (SUBTRACTION)   (MULTIPLICATION)   DIVISION

2. $\dfrac{5d}{6}$

SUM   DIFFERENCE   (PRODUCT)   (QUOTIENT)

3. $5(2 + d) - 8$

   ⬭ADD⬭   ⬭SUBTRACT⬭   ⬭MULTIPLY⬭   DIVIDE

4. abc

   MORE THAN   LESS THAN   ⬭TIMES⬭   EACH

Write an expression using vocabulary words to represent each given expression.

5. $8 - 2g$

   Possible answers: 8 minus the product of 2 and g; 2 times g subtracted from 8; 8 decreased by g doubled

6. $15(a + c)$

   Possible answers: 15 times the quantity of a increased by c; the product of 15 and the sum of a and c; 15 multiplied by the total of a and c

7. $\dfrac{m+n}{5}$

   Possible answers: the sum of m and n divided by 5; the quotient of the total of m and n, and 5; m plus n split into 5 equal groups

8. $b^3 - 18$

   Possible answers: b cubed minus 18; b to the third power decreased by 18

9. $f - \dfrac{d}{2}$

   Possible answers: f minus the quotient of d and 2; d split into 2 groups and then subtracted from f; d divided by 2 less than f

10. $\dfrac{u}{x}$

   Possible answers: u divided by x; the quotient of u and x; u divided into x parts

Closing (5 minutes)

- Peter says the expression $11 - 3c$ is 3 times c decreased by 11. Is he correct? Why or why not?
  - *Peter is not correct because the expression he wrote is in the wrong order. If Peter wanted to write a correct expression and use the same vocabulary words, he would have to write 11 decreased by 3 times c.*

Exit Ticket (5 minutes)

©2015 Great Minds. eureka-math.org
G6-M4-TE-B4-1.3.1-01.2016

Name _____   Date _____

Lesson 15: Read Expressions in Which Letters Stand for Numbers

Exit Ticket

1. Write two word expressions for each problem using different math vocabulary for each expression.

   a. $5d - 10$

   b. $\dfrac{a}{b+2}$

2. List five different math vocabulary words that could be used to describe each given expression.

   a. $3(d - 2) + 10$

   b. $\dfrac{ab}{c}$

EUREKA
MATH™

Exit Ticket Sample Solutions

1. Write two word expressions for each problem using different math vocabulary for each expression.

    a. $5d - 10$

         Possible answers: the product of 5 and d minus 10, 10 less than 5 times d

    b. $\dfrac{a}{b+2}$

         Possible answers: the quotient of a and the quantity of b plus 2, a divided by the sum of b and 2

2. List five different math vocabulary words that could be used to describe each given expression.

    a. $3(d - 2) + 10$

         Possible answers: difference, subtract, product, times, quantity, add, sum

    b. $\dfrac{ab}{c}$

         Possible answers: quotient, divide, split, product, multiply, times, per, each

Problem Set Sample Solutions

1. List five different vocabulary words that could be used to describe each given expression.

    a. $a - d + c$

         Possible answers: sum, add, total, more than, increase, decrease, difference, subtract, less than

    b. $20 - 3c$

         Possible answers: difference, subtract, fewer than, triple, times, product

    c. $\dfrac{b}{d+2}$

         Possible answers: quotient, divide, split, per, sum, add, increase, more than

2. Write an expression using math vocabulary for each expression below.

    a. $5b - 18$

         Possible answers: the product of 5 and b minus 18, 18 less than 5 times b

    b. $\dfrac{n}{2}$

         Possible answers: the quotient of n and 2, n split into 2 equal groups

    c. $a + (d - 6)$

         Possible answers: a plus the quantity d minus 6, a increased by the difference of d and 6

    d. $10 + 2b$

         Possible answers: 10 plus twice b, the total of 10 and the product of 2 and b

Lesson 15:   Read Expressions in Which Letters Stand for Numbers

Lesson 16: Write Expressions in Which Letters Stand for Numbers

Student Outcomes

- Students write algebraic expressions that record all operations with numbers and letters standing for the numbers.

Lesson Notes

In general, key word readings should be avoided. However, at this initial phase, it is important for students to understand the direct relationship between words in a written phrase and their appearance in an algebraic expression.

Classwork

Opening Exercise (5 minutes)

Students underline the key math vocabulary words in each statement.

Opening Exercise

Underline the key words in each statement.

a.   The sum of twice b and 5

   The sum of twice b and 5

b.   The quotient of c and d

   The quotient of c and d

c. a raised to the fifth power and then increased by the product of 5 and c

   a raised to the fifth power and then increased by the product of 5 and c

d.   The quantity of a plus b divided by 4

   The quantity of a plus b divided by 4

e.   10 less than the product of 15 and c

   10 less than the product of 15 and c

f.   5 times d and then increased by 8

   5 times d and then increased by 8

Mathematical Modeling Exercise 1 (10 minutes)

Model how to change the expressions given in the Opening Exercise from words to variables and numbers.

Mathematical Modeling Exercise 1

    a.   The sum of twice b and 5

- Underline key words: the <u>sum</u> of <u>twice</u> b and 5.
- Identify the operations each key word implies.
  - *"Sum" indicates addition, and "twice" indicates multiplication by 2.*
- Write an expression.

    $2b + 5$

    b.   The quotient of c and d

- Underline key words: the <u>quotient</u> of c and d.
- Identify the operation the key word implies.
  - *"Quotient" implies division.*
- Write an expression.

    $\dfrac{c}{d}$

MP.6

    c. a raised to the fifth power and then increased by the product of 5 and c

---

- Underline key words: a raised to the fifth <u>power</u> and then <u>increased</u> by the <u>product</u> of 5 and c.
- Identify the operations each key word implies.
  - *"Power" indicates exponents, "increased" implies addition, and "product" implies multiplication.*
- Write an expression.

    $a^5 + 5c$

    d.   The quantity of a plus b divided by 4

- Underline key words: the <u>quantity</u> of a <u>plus</u> b <u>divided</u> by 4.
- Identify the operations each key word implies.
  - *"Quantity" indicates parentheses, "plus" indicates addition, and "divided by" implies division.*
- Write an expression.

    $\dfrac{a+b}{4}$

    e.   10 less than the product of 15 and c

- Underline key words: 10 <u>less than</u> the <u>product</u> of 15 and c.
- Identify the operations each key word implies.
    - *"Less than" indicates subtraction, and "product" implies multiplication.*
- Write an expression.

> $15c - 10$

MP.6

- Would $10 - 15c$ also be correct? Why or why not?
    - *This expression would not be correct. If the amount of money I have is 10 less than someone else, I would take the money the other person has and subtract the 10.*

> f.   5 times d and then increased by 8

- Underline key words: 5 <u>times</u> d and then <u>increased</u> by 8.
- Identify the operations each key word implies.
    - *"Times" indicates multiplication, and "increased" implies addition.*
- Write an expression.

> $5d + 8$

Mathematical Modeling Exercise 2 (10 minutes)

> **Mathematical Modeling Exercise 2**
>
> Model how to change each real-world scenario to an expression using variables and numbers. Underline the text to show the key words before writing the expression.
>
> Marcus has 4 more dollars than Yaseen. If y is the amount of money Yaseen has, write an expression to show how much money Marcus has.

- Underline key words.
    - *Marcus has 4 <u>more dollars than</u> Yaseen.*

MP.6

- If Yaseen had \$7, how much money would Marcus have?
    - *\$11*
- How did you get that?
    - *Added* $7 + 4$
- Write an expression using y for the amount of money Yaseen has.
    - $y + 4$

> Mario is missing half of his assignments. If a represents the number of assignments, write an expression to show how many assignments Mario is missing.

Lesson 16:   Write Expressions in Which Letters Stand for Numbers

- Underline key words.
  - *Mario is missing <u>half</u> of his assignments.*
- If Mario was assigned 10 assignments, how many is he missing?
  - 5
- How did you get that?
  - $10 \div 2$
- Write an expression using a for the number of assignments Mario was assigned.
  - $\dfrac{a}{2}$ or $a \div 2$

> **Kamilah's weight has tripled since her first birthday. If w represents the amount Kamilah weighed on her first birthday, write an expression to show how much Kamilah weighs now.**

- Underline key words.
  - *Kamilah's weight has <u>tripled</u> since her first birthday.*
- If Kamilah weighed 20 pounds on her first birthday, how much does she weigh now?
  - 60 pounds
- How did you get that?
  - *Multiplied 3 by 20*
- Write an expression using w for Kamilah's weight on her first birthday.
  - $3w$

MP.6

> **Nathan brings cupcakes to school and gives them to his five best friends, who share them equally. If c represents the number of cupcakes Nathan brings to school, write an expression to show how many cupcakes each of his friends receive.**

- Underline key words.
  - *Nathan brings cupcakes to school and gives them to his five best friends, who <u>share</u> them <u>equally</u>.*
- If Nathan brings 15 cupcakes to school, how many will each friend receive?
  - 3
- How did you determine that?
  - $15 \div 5$
- Write an expression using c to represent the number of cupcakes Nathan brings to school.
  - $\dfrac{c}{5}$ or $c \div 5$

> **Mrs. Marcus combines her atlases and dictionaries and then divides them among 10 different tables. If a represents the number of atlases and d represents the number of dictionaries Mrs. Marcus has, write an expression to show how many books would be on each table.**

- Underline key words.
  - *Mrs. Marcus <u>combines</u> her atlases and dictionaries and then <u>divides</u> them among 10 different tables.*

- If Mrs. Marcus had 8 atlases and 12 dictionaries, how many books would be at each table?
  - 2
- How did you determine that?
  - *Added the atlases and dictionaries together and then divided by* 10.
- Write an expression using a for atlases and d for dictionaries to represent how many books each table would receive.
  - $\dfrac{a+d}{10}$ or $(a+d) \div 10$

To improve in basketball, Ivan's coach told him that he needs to take four times as many free throws and four times as many jump shots every day. If f represents the number of free throws and j represents the number of jump shots Ivan shoots daily, write an expression to show how many shots he will need to take in order to improve in basketball.

MP.6

- Underline key words.
  - *To improve in basketball, Ivan needs to shoot* 4 <u>times</u> *more free throws* <u>and</u> *jump shots daily.*
- If Ivan shoots 5 free throws and 10 jump shots, how many will he need to shoot in order to improve in basketball?
  - 60
- How did you determine that?
  - *Added the free throws and jump shots together and then multiplied by* 4
- Write an expression using f for free throws and j for jump shots to represent how many shots Ivan will have to take in order to improve in basketball.
  - $4(f+j)$ or $4f + 4j$

Exercises (10 minutes)

Have students work individually on the following exercises.

Exercises

Mark the text by underlining key words, and then write an expression using variables and/or numbers for each statement.

1. b decreased by c squared

   b <u>decreased</u> by c <u>squared</u>

   $b - c^2$

2. 24 divided by the product of 2 and a

   24 <u>divided</u> by the <u>product</u> of 2 and a

   $\dfrac{24}{2a}$ or $24 \div (2a)$

©2015 Great Minds. eureka-math.org
G6-M4-TE-B4-1.3.1-01.2016

3. 150 decreased by the quantity of 6 plus b

   150 _decreased_ by the _quantity_ of 6 _plus_ b

   $150 - (6 + b)$

4. The sum of twice c and 10

   The _sum_ of _twice_ c and 10

   $2c + 10$

5. Marlo had $\$35$ but then spent $\$m$

   Mario had $\$35$ but then _spent_ $\$m$.

   $35 - m$

6. Samantha saved her money and was able to quadruple the original amount, m.

   Samantha saved her money and was able to _quadruple_ the original amount, m.

   $4m$

7. Veronica increased her grade, g, by 4 points and then doubled it.

   Veronica _increased_ her grade, g, by 4 points and then _doubled_ it.

   $2(g + 4)$

8. Adbell had m pieces of candy and ate 5 of them. Then, he split the remaining candy equally among 4 friends.

   Adbell had m pieces of candy and _ate_ 5 of them. Then, he _split_ the remaining candy equally among 4 friends.

   $\dfrac{m-5}{4}$ or $(m - 5) \div 4$

9. To find out how much paint is needed, Mr. Jones must square the side length, s, of the gate and then subtract 15.

   To find out how much paint is needed, Mr. Jones must _square_ the side length, s, of the gate and then _subtract_ 15.

   $s^2 - 15$

10. Luis brought x cans of cola to the party, Faith brought d cans of cola, and De'Shawn brought h cans of cola. How many cans of cola did they bring altogether?

    Luis _brought_ x cans of cola to the party, Faith _brought_ d cans of cola, _and_ De'Shawn _brought_ h cans of cola. How many cans of cola _did they bring altogether_?

    $x + d + h$

©2015 Great Minds. eureka-math.org
G6-M4-TE-B4-1.3.1-01.2016

Closing (5 minutes)

- How is writing expressions with variables and numbers similar to writing expressions using words?
  - *Possible answers: The same vocabulary words can be used; identifying parts of the expression before writing the expression is helpful.*
- How is writing expressions with variables and numbers different than writing expressions using words?
  - *Possible answers: When an expression with words is provided, it is possible that it might be represented mathematically in more than one way. However, when an algebraic expression is written, there can only be one correct answer.*

Exit Ticket (5 minutes)

Name _____   Date _____

Lesson 16: Write Expressions in Which Letters Stand for Numbers

Exit Ticket

Mark the text by underlining key words, and then write an expression using variables and/or numbers for each of the statements below.

1. Omaya picked x amount of apples, took a break, and then picked v more. Write the expression that models the total number of apples Omaya picked.

2. A number h is tripled and then decreased by 8.

3. Sidney brought s carrots to school and combined them with Jenan's j carrots. She then splits them equally among 8 friends.

4. 15 less than the quotient of e and d

5. Marissa's hair was 10 inches long, and then she cut h inches.

Exit Ticket Sample Solutions

Mark the text by underlining key words, and then write an expression using variables and/or numbers for each of the statements below.

1. Omaya picked x amount of apples, took a break, and then picked v more. Write the expression that models the total number of apples Omaya picked.

   Omaya picked x amount of apples, took a break, and then picked v <u>more</u>.

   $x + v$

2. A number h is tripled and then decreased by 8.

   A number h is <u>tripled</u> and then <u>decreased</u> by 8.

   $3h - 8$

3. Sidney brought s carrots to school and combined them with Jenan's j carrots. She then split them equally among 8 friends.

   Sidney brought s carrots to school and <u>combined</u> them with Jenan's j carrots. She then <u>split</u> them equally among 8 friends.

   $\dfrac{s+j}{8}$ *or* $(s + j) \div 8$

4. 15 less than the quotient of e and d

   15 <u>less than</u> the <u>quotient</u> of e and d

   $\dfrac{e}{d} - 15$ *or* $e \div d - 15$

5. Marissa's hair was 10 inches long, and then she cut h inches.

   Marissa's hair was 10 inches long, and then she <u>cut</u> h inches.

   $10 - h$

Problem Set Sample Solutions

Mark the text by underlining key words, and then write an expression using variables and numbers for each of the statements below.

1. Justin can type w words per minute. Melvin can type 4 times as many words as Justin. Write an expression that represents the rate at which Melvin can type.

   Justin can type w words per minute. Melvin can type 4 <u>times</u> as many words as Justin. Write an expression that represents the rate at which Melvin can type.

   $4w$

2.   Yohanna swam y yards yesterday. Sheylin swam 5 yards less than half the amount of yards as Yohanna. Write an expression that represents the number of yards Sheylin swam yesterday.

Yohanna swam y yards yesterday. Sheylin swam 5 yards <u>less than half</u> the amount of yards as Yohanna. Write an expression that represents the number of yards Sheylin swam yesterday.

$\frac{y}{2} - 5 \text{ or } y \div 2 - 5 \text{ or } \frac{1}{2}y - 5$

3.   A number d is decreased by 5 and then doubled.

A number d is <u>decreased by</u> 5 and then <u>doubled</u>.

$2(d - 5)$

4.   Nahom had n baseball cards, and Semir had s baseball cards. They combined their baseball cards and then sold 10 of them.

Nahom had n baseball cards, and Semir had s baseball cards. They <u>combined</u> their baseball cards and then <u>sold</u> 10 of them.

$n + s - 10$

5.   The sum of 25 and h is divided by f cubed.

The <u>sum</u> of 25 and h is <u>divided</u> by f <u>cubed</u>.

$\frac{25+h}{f^3} \text{ or } (25 + h) \div f^3$

Lesson 17: Write Expressions in Which Letters Stand for Numbers

Student Outcomes

- Students write algebraic expressions that record all operations with numbers and/or letters standing for the numbers.

Lesson Notes

Large paper is needed to complete this lesson.

Classwork

Fluency Exercise (5 minutes): Addition of Decimals

Sprint: Refer to the Sprints and Sprint Delivery Script sections in the Module Overview for directions on how to administer a Sprint.

Opening (5 minutes)

Discuss the Exit Ticket from Lesson 16. Students continue to work on writing expressions, so discuss any common mistakes from the previous lesson.

Exercises (25 minutes)

Students work in groups of two or three to complete the stations. At each station, students write down the problem and the expression with variables and/or numbers. Encourage students to underline key words in each problem.

Exercises

Station One:

1. The sum of a and b

    $a + b$

2. Five more than twice a number c

    $5 + 2c \text{ or } 2c + 5$

3. Martha bought d number of apples and then ate 6 of them.

    $d - 6$

> *Scaffolding:*
>
> If students struggled during Lesson 16, complete some examples with students before moving into the exercises.

Station Two:

1. 14 decreased by p

   $14 - p$

2. The total of d and f, divided by 8

   $\dfrac{d+f}{8}$ or $(d + f) \div 8$

3. Rashod scored 6 less than 3 times as many baskets as Mike. Mike scored b baskets.

   $3b - 6$

Station Three:

1. The quotient of c and 6

   $\dfrac{c}{6}$

2. Triple the sum of x and 17

   $3(x + 17)$

3. Gabrielle had b buttons but then lost 6. Gabrielle took the remaining buttons and split them equally among her 5 friends.

   $\dfrac{b-6}{5}$ or $(b - 6) \div 5$

Station Four:

1. d doubled

   $2d$

2. Three more than 4 times a number x

   $4x + 3$ or $3 + 4x$

3. Mali has c pieces of candy. She doubles the amount of candy she has and then gives away 15 pieces.

   $2c - 15$

Station Five:

1. f cubed

   f^3

2. The quantity of 4 increased by a, and then the sum is divided by 9.

   $\dfrac{4+a}{9}$ or $(4 + a) \div 9$

3. Tai earned 4 points fewer than double Oden's points. Oden earned p points.

   $2p - 4$

Lesson 17:   Write Expressions in Which Letters Stand for Numbers

171

©2015 Great Minds. eureka-math.org
G6-M4-TE-B4-1.3.1-01.2016

Station Six:

1. The difference between d and 8

    $d - 8$

2. 6 less than the sum of d and 9

    $(d + 9) - 6$

3. Adalyn has x pants and s shirts. She combined them and sold half of them. How many items did Adalyn sell?

    $\frac{x+s}{2}$ or $\frac{1}{2}(x + s)$

When students reach the final station, they complete the station on larger paper. Students should put all of their work on the top half of the paper.

MP.3

After all students have completed every station, they travel through the stations again to look at the answers provided on the larger paper. Students compare their answers with the answers at the stations and leave feedback on the bottom half of the paper. This may be positive feedback ("I agree with all of your answers" or "Great job") or critiques ("I think your subtraction is in the incorrect order" or "Why did you write your answer in that order?").

Closing (5 minutes)

Discuss feedback that was left on the larger sheets of paper. This answers any questions and provides an opportunity to discuss common mistakes.

- Is it possible to have more than one correct answer? Why or why not?

    □ *When writing some of the expressions, it is possible to have more than one correct answer. For example, when writing an expression with addition, the order can be different. Also, we learned how to write division expressions in more than one way.*

Exit Ticket (5 minutes)

Name _____   Date _____

Lesson 17: Write Expressions in Which Letters Stand for Numbers

Exit Ticket

Write an expression using letters and/or numbers for each problem below.

1. d squared

2. A number x increased by 6, and then the sum is doubled.

3. The total of h and b is split into 5 equal groups.

4. Jazmin has increased her $45 by m dollars and then spends a third of the entire amount.

5. Bill has d more than 3 times the number of baseball cards as Frank. Frank has f baseball cards.

Exit Ticket Sample Solutions

Write an expression using letters and/or numbers for each problem below.

1. d squared

   d^2

2. A number x increased by 6, and then the sum is doubled.

   $2(x + 6)$

3. The total of h and b is split into 5 equal groups.

   $\frac{h+b}{5}$ or $(h + b) \div 5$

4. Jazmin has increased her \$45 by m dollars and then spends a third of the entire amount.

   $\frac{45 + m}{3}$ or $\frac{1}{3}(45 + m)$

5. Bill has d more than 3 times the number of baseball cards as Frank. Frank has f baseball cards.

   $3f + d$ or $d + 3f$

Problem Set Sample Solutions

Write an expression using letters and/or numbers for each problem below.

1. 4 less than the quantity of 8 times n

   $8n - 4$

2. 6 times the sum of y and 11

   $6(y + 11)$

3. The square of m reduced by 49

   $m^2 - 49$

4. The quotient when the quantity of 17 plus p is divided by 8

   $\frac{17+p}{8}$ or $(17 + p) \div 8$

5. Jim earned j in tips, and Steve earned s in tips. They combine their tips and then split them equally.

   $\frac{j+s}{2}$ or $(j + s) \div 2$

6. Owen has c collector cards. He quadruples the number of cards he has and then combines them with Ian, who has i collector cards.

$4c + i$

7. Rae runs 4 times as many miles as Madison and Aaliyah combined. Madison runs m miles, and Aaliyah runs a miles.

$4(m + a)$

8. By using coupons, Mary Jo is able to decrease the retail price of her groceries, g, by $\$125$.

$g - 125$

9. To calculate the area of a triangle, you find the product of the base and height and then divide by 2.

$\dfrac{bh}{2}$ or $bh \div 2$

10. The temperature today was 10 degrees colder than twice yesterday's temperature, t.

$2t - 10$

Number Correct: _____

Addition of Decimals I—Round 1

Directions: Evaluate each expression.

1.	5.1 + 6		23.	3.6 + 2.1	
2.	5.1 + 0.6		24.	3.6 + 0.21	
3.	5.1 + 0.06		25.	3.6 + 0.021	
4.	5.1 + 0.006		26.	0.36 + 0.021	
5.	5.1 + 0.0006		27.	0.036 + 0.021	
6.	3 + 2.4		28.	1.4 + 42	
7.	0.3 + 2.4		29.	1.4 + 4.2	
8.	0.03 + 2.4		30.	1.4 + 0.42	
9.	0.003 + 2.4		31.	1.4 + 0.042	
10.	0.0003 + 2.4		32.	0.14 + 0.042	
11.	24 + 0.3		33.	0.014 + 0.042	
12.	2 + 0.3		34.	0.8 + 2	
13.	0.2 + 0.03		35.	0.8 + 0.2	
14.	0.02 + 0.3		36.	0.08 + 0.02	
15.	0.2 + 3		37.	0.008 + 0.002	
16.	2 + 0.03		38.	6 + 0.4	
17.	5 + 0.4		39.	0.6 + 0.4	
18.	0.5 + 0.04		40.	0.06 + 0.04	
19.	0.05 + 0.4		41.	0.006 + 0.004	
20.	0.5 + 4		42.	0.1 + 9	
21.	5 + 0.04		43.	0.1 + 0.9	
22.	0.5 + 0.4		44.	0.01 + 0.09	

Lesson 17:   Write Expressions in Which Letters Stand for Numbers

EUREKA MATH

Addition of Decimals I—Round 1 [KEY]

Directions: Evaluate each expression.

1.	5.1 + 6	**11.1**	23.	3.6 + 2.1	**5.7**
2.	5.1 + 0.6	**5.7**	24.	3.6 + 0.21	**3.81**
3.	5.1 + 0.06	**5.16**	25.	3.6 + 0.021	**3.621**
4.	5.1 + 0.006	**5.106**	26.	0.36 + 0.021	**0.381**
5.	5.1 + 0.0006	**5.1006**	27.	0.036 + 0.021	**0.057**
6.	3 + 2.4	**5.4**	28.	1.4 + 42	**43.4**
7.	0.3 + 2.4	**2.7**	29.	1.4 + 4.2	**5.6**
8.	0.03 + 2.4	**2.43**	30.	1.4 + 0.42	**1.82**
9.	0.003 + 2.4	**2.403**	31.	1.4 + 0.042	**1.442**
10.	0.0003 + 2.4	**2.4003**	32.	0.14 + 0.042	**0.182**
11.	24 + 0.3	**24.3**	33.	0.014 + 0.042	**0.056**
12.	2 + 0.3	**2.3**	34.	0.8 + 2	**2.8**
13.	0.2 + 0.03	**0.23**	35.	0.8 + 0.2	**1**
14.	0.02 + 0.3	**0.32**	36.	0.08 + 0.02	**0.1**
15.	0.2 + 3	**3.2**	37.	0.008 + 0.002	**0.01**
16.	2 + 0.03	**2.03**	38.	6 + 0.4	**6.4**
17.	5 + 0.4	**5.4**	39.	0.6 + 0.4	**1**
18.	0.5 + 0.04	**0.54**	40.	0.06 + 0.04	**0.1**
19.	0.05 + 0.4	**0.45**	41.	0.006 + 0.004	**0.01**
20.	0.5 + 4	**4.5**	42.	0.1 + 9	**9.1**
21.	5 + 0.04	**5.04**	43.	0.1 + 0.9	**1**
22.	0.5 + 0.4	**0.9**	44.	0.01 + 0.09	**0.1**

EUREKA MATH™

Lesson 17:   Write Expressions in Which Letters Stand for Numbers

177

©2015 Great Minds. eureka-math.org
G6-M4-TE-B4-1.3.1-01.2016

Number Correct: _____
Improvement: _____

Addition of Decimals I—Round 2

Directions: Evaluate each expression.

1.	3.2 + 5		23.	4.2 + 5.5	
2.	3.2 + 0.5		24.	4.2 + 0.55	
3.	3.2 + 0.05		25.	4.2 + 0.055	
4.	3.2 + 0.005		26.	0.42 + 0.055	
5.	3.2 + 0.0005		27.	0.042 + 0.055	
6.	4 + 5.3		28.	2.7 + 12	
7.	0.4 + 5.3		29.	2.7 + 1.2	
8.	0.04 + 5.3		30.	2.7 + 0.12	
9.	0.004 + 5.3		31.	2.7 + 0.012	
10.	0.0004 + 5.3		32.	0.27 + 0.012	
11.	4 + 0.53		33.	0.027 + 0.012	
12.	6 + 0.2		34.	0.7 + 3	
13.	0.6 + 0.02		35.	0.7 + 0.3	
14.	0.06 + 0.2		36.	0.07 + 0.03	
15.	0.6 + 2		37.	0.007 + 0.003	
16.	2 + 0.06		38.	5 + 0.5	
17.	1 + 0.7		39.	0.5 + 0.5	
18.	0.1 + 0.07		40.	0.05 + 0.05	
19.	0.01 + 0.7		41.	0.005 + 0.005	
20.	0.1 + 7		42.	0.2 + 8	
21.	1 + 0.07		43.	0.2 + 0.8	
22.	0.1 + 0.7		44.	0.02 + 0.08	

EUREKA MATH™

Addition of Decimals I—Round 2 [KEY]

Directions: Evaluate each expression.

1.	3.2 + 5	**8.2**	23.	4.2 + 5.5	**9.7**
2.	3.2 + 0.5	**3.7**	24.	4.2 + 0.55	**4.75**
3.	3.2 + 0.05	**3.25**	25.	4.2 + 0.055	**4.255**
4.	3.2 + 0.005	**3.205**	26.	0.42 + 0.055	**0.475**
5.	3.2 + 0.0005	**3.2005**	27.	0.042 + 0.055	**0.097**
6.	4 + 5.3	**9.3**	28.	2.7 + 12	**14.7**
7.	0.4 + 5.3	**5.7**	29.	2.7 + 1.2	**3.9**
8.	0.04 + 5.3	**5.34**	30.	2.7 + 0.12	**2.82**
9.	0.004 + 5.3	**5.304**	31.	2.7 + 0.012	**2.712**
10.	0.0004 + 5.3	**5.3004**	32.	0.27 + 0.012	**0.282**
11.	4 + 0.53	**4.53**	33.	0.027 + 0.012	**0.039**
12.	6 + 0.2	**6.2**	34.	0.7 + 3	**3.7**
13.	0.6 + 0.02	**0.62**	35.	0.7 + 0.3	**1**
14.	0.06 + 0.2	**0.26**	36.	0.07 + 0.03	**0.1**
15.	0.6 + 2	**2.6**	37.	0.007 + 0.003	**0.01**
16.	2 + 0.06	**2.06**	38.	5 + 0.5	**5.5**
17.	1 + 0.7	**1.7**	39.	0.5 + 0.5	**1**
18.	0.1 + 0.07	**0.17**	40.	0.05 + 0.05	**0.1**
19.	0.01 + 0.7	**0.71**	41.	0.005 + 0.005	**0.01**
20.	0.1 + 7	**7.1**	42.	0.2 + 8	**8.2**
21.	1 + 0.07	**1.07**	43.	0.2 + 0.8	**1**
22.	0.1 + 0.7	**0.8**	44.	0.02 + 0.08	**0.1**

Lesson 17:   Write Expressions in Which Letters Stand for Numbers

©2015 Great Minds. eureka-math.org
G6-M4-TE-B4-1.3.1-01.2016

Name _____   Date _____

1. Yolanda is planning out her vegetable garden. She decides that her garden will be square. Below are possible sizes of the garden she will create.

   a. Complete the table by continuing the pattern.

Side Length	1 foot	2 feet	3 feet	4 feet	5 feet	x feet
Notation	$1^2 = 1 \cdot 1 = 1$					
Formula	$A = l \cdot w$ $A = 1 \text{ ft} \cdot 1 \text{ ft}$ $A = 1^2 \text{ ft}^2$ $A = 1 \text{ ft}^2$					
Representation	☐	⊞				

   b. Yolanda decides the length of her square vegetable garden will be 17 ft. She calculates that the area of the garden is 34 ft². Determine if Yolanda's calculation is correct. Explain.

©2015 Great Minds. eureka-math.org
G6-M4-TE-B4-1.3.1-01.2016

EUREKA
MATH™

2. Yolanda creates garden cubes to plant flowers. She will fill the cubes with soil and needs to know the amount of soil that will fill each garden cube. The volume of a cube is determined by the following formula: $V = s^3$, where s represents the side length.

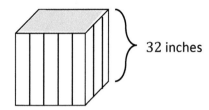

32 inches

     a. Represent the volume, in cubic inches, of the garden cube above using a numerical expression.

     b. Evaluate the expression to determine the volume of the garden cube and the amount of soil, in cubic inches, she will need for each cube.

3. Explain why $\left(\dfrac{1}{2}\right)^4 = \dfrac{1}{16}$.

4. Yolanda is building a patio in her backyard. She is interested in using both brick and wood for the flooring of the patio. Below is the plan she has created for the patio. All measurements are in feet.

   a. Create an expression to represent the area of the patio.

   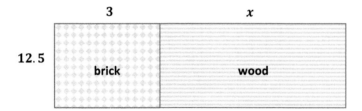

   b. Yolanda's husband develops another plan for the patio because he prefers the patio to be much wider than Yolanda's plan. Determine the length of the brick section and the length of the wood section. Then, use the dimensions to write an expression that represents the area of the entire patio.

   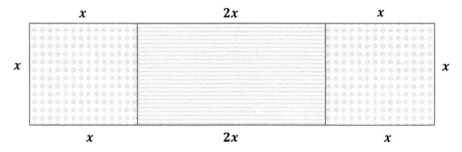

5. The landscaper hired for Yolanda's lawn suggests a patio that has the same measure of wood as it has brick.

   a. Express the perimeter of the patio in terms of x, first using addition and then using multiplication.

   b. Use substitution to determine if your expressions are equivalent. Explain.

6. Elena and Jorge have similar problems and find the same answer. Each determines that the solution to the problem is 24.

Elena: $(14 + 42) \div 7 + 4^2$ Jorge: $14 + (42 \div 7) + 4^2$

a. Evaluate each expression to determine if both Elena and Jorge are correct.

b. Why would each find the solution of 24? What mistakes were made, if any?

7. Jackson gave Lena this expression to evaluate: $14(8 + 12)$. Lena said that to evaluate the expression was simple; just multiply the factors 14 and 20. Jackson told Lena she was wrong. He solved it by finding the product of 14 and 8 and then adding that to the product of 14 and 12.

a. Evaluate the expression using each student's method.

Lena's Method	Jackson's Method

b. Who was right in this discussion? Why?

A Progression Toward Mastery					
Assessment Task Item		**STEP 1** Missing or incorrect answer and little evidence of reasoning or application of mathematics to solve the problem.	**STEP 2** Missing or incorrect answer but evidence of some reasoning or application of mathematics to solve the problem.	**STEP 3** A correct answer with some evidence of reasoning or application of mathematics to solve the problem, OR an incorrect answer with substantial evidence of solid reasoning or application of mathematics to solve the problem.	**STEP 4** A correct answer supported by substantial evidence of solid reasoning or application of mathematics to solve the problem.
1	a 6.EE.A.1 6.EE.A.2	Student completes the table with fewer than 10 cells correct. An entire row may be incorrect, indicating that incomplete understanding is still present.	Student completes the table with at least 10 of the 15 cells correct. One or more of the three rows in the last column are incorrect, indicating the student does not understand the general form of the notation, formula, and representation.	Student completes the table with at least 13 of the 15 cells correct. All three rows in the last column are correct, indicating the student understands the general form of the notation, formula, and representation.	Student completes the table without error. All notations, formulas, and representations are correct. Please note student exemplar below.

Side Length	1 foot	2 feet	3 feet	4 feet	5 feet	x feet
Notation	$1^2 = 1 \cdot 1 = 1$	$2^2 = 2 \cdot 2 = 4$	$3^2 = 3 \cdot 3 = 9$	$4^2 = 4 \cdot 4 = 16$	$5^2 = 5 \cdot 5 = 25$	$x^2 = x \cdot x = x^2$
Formula	$A = l \cdot w$ $A = 1\,\text{ft} \cdot 1\,\text{ft}$ $A = 1^2\,\text{ft}^2$ $A = 1\,\text{ft}^2$	$A = l \cdot w$ $A = 2\,\text{ft} \cdot 2\,\text{ft}$ $A = 2^2\,\text{ft}^2$ $A = 4\,\text{ft}^2$	$A = l \cdot w$ $A = 3\,\text{ft} \cdot 3\,\text{ft}$ $A = 3^2\,\text{ft}^2$ $A = 9\,\text{ft}^2$	$A = l \cdot w$ $A = 4\,\text{ft} \cdot 4\,\text{ft}$ $A = 4^2\,\text{ft}^2$ $A = 16\,\text{ft}^2$	$A = l \cdot w$ $A = 5\,\text{ft} \cdot 5\,\text{ft}$ $A = 5^2\,\text{ft}^2$ $A = 25\,\text{ft}^2$	$A = l \cdot w$ $A = x\,\text{ft} \cdot x\,\text{ft}$ $A = x^2\,\text{ft}^2$
Representation	▢	⊞	(3×3 grid)	(4×4 grid)	(5×5 grid)	(square labeled x)

| | b

6.EE.A.1
6.EE.A.2 | Student states that Yolanda was correct in her calculation or states that Yolanda was incorrect but offers no explanation. | Student states that Yolanda was incorrect but offers an incomplete analysis of the error. | Student states that Yolanda was incorrect and that Yolanda calculated $17 \cdot 2$. | Student states that Yolanda was incorrect and that Yolanda found $17 \cdot 2$, the base times the exponent. AND Student calculates the correct area: $17^2\,\text{ft}^2 = 289\,\text{ft}^2$. |

Module 4:   Expressions and Equations

2	**a** 6.EE.A.2	Student does not write a numerical expression or writes an expression unrelated to the problem.	Student writes a numerical expression that relates the volume and side length, but the student makes an error, such as using 2 as an exponent.	Student writes an equation that correctly represents the data, $V = 32^3$ or $V = 32 \cdot 32 \cdot 32$, instead of a numerical expression.	Student correctly writes the numerical expression for the volume of the cube: 32^3, or $32 \cdot 32 \cdot 32$.
	b 6.EE.A.1	Student does not attempt to evaluate the expression or has no expression from part (a) to evaluate.	Student attempts to evaluate the expression but makes an arithmetic error.	Student correctly evaluates the expression and finds $V = 32768$. The unit of volume, in^3, is missing. OR Student correctly evaluates an incorrect expression from part (a).	Student correctly evaluates the expression and uses the correct unit. Student gives the answer $V = 32{,}768$ in^3.
3	6.EE.A.1	Student does not demonstrate understanding of exponential notation. One example would be adding $\frac{1}{2}$ four times.	Student makes a common error, such as $\left(\frac{1}{2}\right)^4 = \frac{4}{2}$ or $\left(\frac{1}{2}\right)^4 = \frac{1}{8}$.	Student shows that $\left(\frac{1}{2}\right)^4 = \frac{1}{2} \cdot \frac{1}{2} \cdot \frac{1}{2} \cdot \frac{1}{2}$ but makes an arithmetic error and arrives at an answer other than $\frac{1}{16}$.	Student shows that $\left(\frac{1}{2}\right)^4 = \frac{1}{2} \cdot \frac{1}{2} \cdot \frac{1}{2} \cdot \frac{1}{2} = \frac{1}{16}$.
4	**a** 6.EE.A.3	Student does not write an expression or does not indicate an understanding of $A = l \cdot w$.	Student writes an expression relating the width (12.5) to only one part of the length (3 or x).	Student writes the expression incorrectly, without parentheses: $12.5 \cdot 3 + x$, but includes each term needed to find the area.	Student writes the correct expression: $12.5(3 + x)$ or $37.5 + 12.5x$.
	b 6.EE.A.3	Student does not write an expression or does not indicate an understanding of $A = l \cdot w$.	Student writes an expression using the width, 24 feet, but does not calculate the length, $2x + 4$, correctly.	Student writes the correct expression, $24(2x + 4)$.	Student writes the correct expression, $24(2x + 4)$, and identifies the width, 24 feet, and the length, $2x + 4$ feet.
5	**a** 6.EE.A.3	Student does not express the perimeter of the figure in terms of x, using neither addition nor multiplication.	Student expresses the perimeter of the figure in terms of x, using either addition or multiplication but not both.	Student expresses the perimeter of the figure in terms of x, using addition and multiplication but makes an error in one of the expressions.	Student expresses the perimeter of the figure as: $x + x + x + 2x + 2x + x + x + x$ (or uses any other order of addends that is equivalent) and writes the expression $10x$.

	b **6.EE.A.4**	Student states that the expressions are not equivalent, but student does not use substitution and offers no explanation.	Student states that the expressions are equivalent, but student does not use substitution and offers no explanation.	Student substitutes a value for x in both equations but makes one or more arithmetic mistakes and claims that the two expressions are not equivalent.	Student substitutes any value for x into both the addition and multiplication expression, calculates them accurately, and finds them equivalent.
6	**a** **6.EE.A.1** **6.EE.A.2**	Student evaluates both expressions incorrectly. Errors are both in order of operations and arithmetic.	Student evaluates one expression correctly and one incorrectly. Errors are due to lack of application of order of operation rules.	Student follows the correct order of operations on both expressions but fails to compute the exponents correctly on one or both expressions.	Student evaluates both expressions accurately, applying the rules of order of operations correctly. Elena's answer is 24, and Jorge's answer is 36.
	b **6.EE.A.1** **6.EE.A.2**	Student offers no credible reason why both Elena and Jorge would arrive at the answer, 24. Jorge's mistakes are not identified.	Student shows partial understanding of order of operations but is unable to find or describe Jorge's mistake. Student may have an incomplete understanding of exponents.	Student finds that Elena followed the order of operation rules correctly. Jorge's mistake is noted, but it is not described in detail.	Student finds that Elena followed the order of operation rules correctly. Also, Jorge's mistake is identified: Jorge did not evaluate the operation inside the parentheses first. Instead, he added $14 + 42$ first, arriving at a sum of 56. He then divided 56 by 7 to get 8, added 4^2 to 8, and arrived at a final answer of 24.
7	**a** **6.EE.A.3**	Student evaluates neither Lena's nor Jackson's methods accurately.	Student evaluates both expressions using the same method.	Student evaluates either Lena's or Jackson's methods accurately. The other is evaluated inaccurately.	Student evaluates both Lena's and Jackson's methods accurately.

Module 4:   Expressions and Equations

b 6.EE.A.2 6.EE.A.3	Student claims both Lena and Jackson are incorrect. Evidence is missing or lacking.	Student chooses either Lena or Jackson as being correct, implying that the other is wrong. Evidence is not fully articulated.	Student indicates that methods used by both Lena and Jackson are correct. Both methods are described. No mention of the distributive property is made.	Student indicates that methods used by both Lena and Jackson are correct. Student claims Lena followed the order of operations by adding $8 + 12$ first because they were contained in parentheses, and then Lena multiplied the sum (20) by 14 to arrive at a product of 280. Student also identifies Jackson's method as an application of the distributive property. Partial products of $14(8) = 112$ and $14(12) = 168$ are found first and then added to arrive at 280.

Name _____   Date _____

1. Yolanda is planning out her vegetable garden. She decides that her garden will be square. Below are possible sizes of the garden she will create.

   a. Complete the table by continuing the pattern.

Side Length	1 foot	2 feet	3 feet	4 feet	5 feet	x feet
Notation	$1^2 = 1 \cdot 1 = 1$	$2^2 = 2 \cdot 2 = 4$	$3^2 = 3 \cdot 3 = 9$	$4^2 = 4 \cdot 4 = 16$	$5^2 = 5 \cdot 5 = 25$	$x^2 = x \cdot x = x^2$
Formula	$A = l \cdot w$ $A = 1\text{ ft} \cdot 1\text{ ft}$ $A = 1^2 \text{ ft}^2$ $A = 1 \text{ ft}^2$	$A = l \cdot w$ $A = 2\text{ft} \cdot 2\text{ft}$ $A = 2^2 \text{ft}^2$ $A = 4 \text{ft}^2$	$A = l \cdot w$ $A = 3\text{ft} \cdot 3\text{ft}$ $A = 3^2 \text{ft}^2$ $A = 9 \text{ft}^2$	$A = l \cdot w$ $A = 4\text{ft} \cdot 4\text{ft}$ $A = 4^2 \text{ft}^2$ $A = 16 \text{ft}^2$	$A = l \cdot w$ $A = 5\text{ft} \cdot 5\text{ft}$ $A = 5^2 \text{ft}^2$ $A = 25 \text{ft}^2$	$A = l \cdot w$ $A = x\text{ ft} \cdot x\text{ ft}$ $A = x^2 f^2$
Representation						

   b. Yolanda decides the length of her square vegetable garden will be 17 ft. She calculates that the area of the garden is 34 ft². Determine if Yolanda's calculation is correct. Explain.

   $A = l \cdot w$
   $A = 17\text{ft} \cdot 17\text{ft}$
   $A = 17^2 \text{ft}^2$
   $A = 289 \text{ft}^2$

   Yolanda is incorrect. Instead of finding 17^2 (or $17 \cdot 17$), she multiplied $17 \cdot 2$, the base times the exponent.

Module 4:   Expressions and Equations

EUREKA MATH™

2. Yolanda creates garden cubes to plant flowers. She will fill the cubes with soil and needs to know the amount of soil that will fill each garden cube. The volume of a cube is determined by the following formula: $V = s^3$, where s equals the side length.

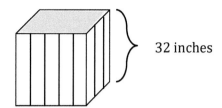

32 inches

a. Represent the volume, in cubic inches, of the garden cube above using a numerical expression.

$$32^3 \text{ or } 32 \cdot 32 \cdot 32$$

b. Evaluate the expression to determine the volume of the garden cube and the amount of soil, in cubic inches, she will need for each cube.

$$32^3 =$$
$$32 \cdot 32 \cdot 32 =$$
$$32,768$$

3. Explain why $\left(\frac{1}{2}\right)^4 = \frac{1}{16}$.

$$\left(\frac{1}{2}\right)^4 = \frac{1}{2} \cdot \frac{1}{2} \cdot \frac{1}{2} \cdot \frac{1}{2} = \frac{1}{16}$$
$$\qquad\qquad 4 \quad 8 \quad 16$$

4. Yolanda is building a patio in her backyard. She is interested in using both brick and wood for the flooring of the patio. Below is the plan she has created for the patio. All measurements are in feet.

a. Create an expression to represent the area of the patio.

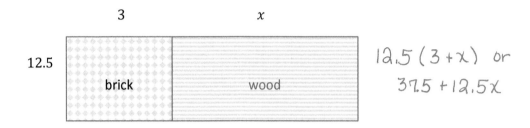

$12.5(3+x)$ or

$37.5 + 12.5x$

b. Yolanda's husband develops another plan for the patio because he prefers the patio to be much wider than Yolanda's plan. Determine the length of the brick section and the length of the wood section. Then, use the dimensions to write an expression that represents the area of the entire patio.

$24(2x+4)$

5. The landscaper hired for Yolanda's lawn suggests a patio that has the same measure of wood as it has brick.

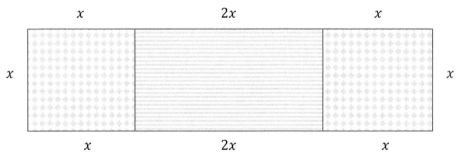

a. Express the perimeter of the patio in terms of x, first using addition and then using multiplication.

$x + x + x + 2x + 2x + x + x + x$

$10x$

b. Use substitution to determine if your expressions are equivalent. Explain.

These expressions are equivalent. Let $x = 2$.

$x + x + x + 2x + 2x + x + x + x$

$2 + 2 + 2 + 4 + 4 + 2 + 2 + 2 = 20$

or

$10x$

$10(2) =$

20

EUREKA
MATH™

6. Elena and Jorge have similar problems and find the same answer. Each determines that the solution to the problem is 24.

Elena: $(14 + 42) \div 7 + 4^2$

$$56 \div 7 + 4^2 =$$
$$56 \div 7 + 16 =$$
$$8 + 16 =$$
$$24$$

Jorge: $14 + (42 \div 7) + 4^2$

$$14 + 6 + 4^2 =$$
$$14 + 6 + 16 =$$
$$20 + 16 =$$
$$36$$

a. Evaluate each expression to determine if both Elena and Jorge are correct.

b. Why would each find the solution of 24? What mistakes were made, if any?

Elena followed the order of operations correctly. Jorge made a mistake. He added 14 + 42 first, and then divided the sum by 7 to get 8. He did not follow the correct order of operations. He should have evaluated the parentheses first.

7. Jackson gave Lena this expression to evaluate: $14(8 + 12)$. Lena said that to evaluate the expression was simple; just multiply the factors 14 and 20. Jackson told Lena she was wrong. He solved it by finding the product of 14 and 8 and then adding that to the product of 14 and 12.

a. Evaluate the expression using each student's method.

Lena's Method	Jackson's Method
$14(8+12) =$ $14(20) =$ 280	$14(8+12)$ $112 + 168 =$ 280

b. Who was right in this discussion? Why?

They were both correct. Lena used the order of operations correctly to determine 280. Jackson used the distributive property correctly to determine 280.

6 GRADE

Mathematics Curriculum

Topic F

Writing and Evaluating Expressions and Formulas

6.EE.A.2a, 6.EE.A.2c, 6.EE.B.6

Focus Standards:	6.EE.A.2	Write, read, and evaluate expressions in which letters stand for numbers.
		a. Write expressions that record operations with numbers and with letters standing for numbers. *For example, express the calculation "Subtract y from 5" as $5 - y$.*
		b. Evaluate expressions at specific values of their variables. Include expressions that arise from formulas used in real-world problems. Perform arithmetic operations, including those involving whole-number exponents, in the conventional order when there are no parentheses to specify a particular order (Order of Operations). *For example, use the formulas $V = s^3$ and $A = 6s^2$ to find the volume and surface area of a cube with sides of length $s = 1/2$.*
	6.EE.B.6	Use variables to represent numbers and write expressions when solving a real-world or mathematical problem; understand that a variable can represent an unknown number, or, depending on the purpose at hand, any number in a specific set.
Instructional Days:	5	
Lesson 18:		Writing and Evaluating Expressions—Addition and Subtraction (P)[1]
Lesson 19:		Substituting to Evaluate Addition and Subtraction Expressions (P)
Lesson 20:		Writing and Evaluating Expressions—Multiplication and Division (P)
Lesson 21:		Writing and Evaluating Expressions—Multiplication and Addition (M)
Lesson 22:		Writing and Evaluating Expressions—Exponents (P)

---

[1]Lesson Structure Key: **P**-Problem Set Lesson, **M**-Modeling Cycle Lesson, **E**-Exploration Lesson, **S**-Socratic Lesson

In Topic F, students demonstrate their knowledge of expressions from previous topics in order to write and evaluate expressions and formulas. Students bridge their understanding of reading and writing expressions to substituting values in order to evaluate expressions. In Lesson 18, students use variables to write expressions involving addition and subtraction from real-world problems. They evaluate those expressions when they are given the value of the variable. For example, given the problem "Quentin has two more dollars than his sister, Juanita," students determine the variable to represent the unknown. In this case, students let x represent Juanita's money, in dollars. Since Quentin has two more dollars than Juanita, students represent his quantity as $x + 2$. Now students can substitute given values for the variable to determine the amount of money Quentin and Juanita each have. If Juanita has fourteen dollars, students substitute the x with the amount, 14, and evaluate the expression: $x + 2$.

$$14 + 2$$
$$16$$

Here, students determine that the amount of money Quentin has is 16 dollars because 16 is two more than the 14 dollars Juanita has.

In Lesson 19, students develop expressions involving addition and subtraction from real-world problems. They use tables to organize the information provided and evaluate expressions for given values. They continue to Lesson 20 where they develop expressions again, this time focusing on multiplication and division from real-world problems. Students bridge their study of the relationships between operations from Topic A to further develop and evaluate expressions in Lesson 21, focusing on multiplication and addition in real-world contexts.

Building from their previous experiences in this topic, students create formulas in Lesson 22 by setting expressions equal to another variable. Students assume, for example, that there are p peanuts in a bag. There are three bags and four extra peanuts altogether. Students express the total number of peanuts in terms of p: $3p + 4$. Students let t be the total number of peanuts and determine a formula that expresses the relationship between the number of peanuts in a bag and the total number of peanuts, $t = 3p + 4$. From there, students are provided a value for p, which they substitute into the formula: If $p = 10$, then $3(10) + 4 = 30 + 4 = 34$, and they determine that there are 34 peanuts.

In the final lesson of the topic, students evaluate formulas involving exponents for given values in real-world problems.

Lesson 18: Writing and Evaluating Expressions—Addition and Subtraction

Student Outcomes

- Students use variables to write expressions involving addition and subtraction from real-world problems.
- Students evaluate these expressions when given the value of the variable.

Lesson Notes

When students write expressions, make sure they are as specific as possible. Students should understand the importance of specifying units when defining letters. For example, students should say, "Let K represent Karolyn's weight in pounds" instead of "Let K represent Karolyn's weight" because weight is not a number until it is specified by pounds, ounces, grams, and so on. They also must be taught that it is inaccurate to define K as Karolyn because Karolyn is not a number. Students conclude that in word problems, each letter represents a number, and its meaning must be clearly stated.

Classwork

Opening Exercise (4 minutes)

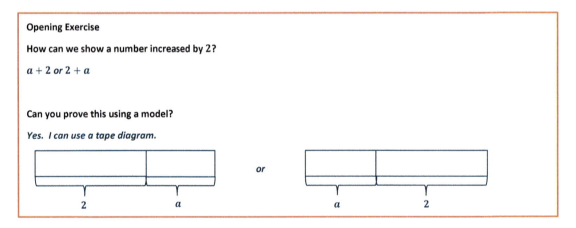

Opening Exercise

How can we show a number increased by 2?

$a + 2$ or $2 + a$

Can you prove this using a model?

Yes. I can use a tape diagram.

2   a   or   a   2

Discussion (5 minutes)

- In this lesson, you connect real-world problems to addition and subtraction expressions. What story problem could you make up to go along with the expression $a + 2$?

Allow a few moments for students to form realistic scenarios. As students share these, critique them.

- *Answers will vary. Ronnie has some apples, but Gayle has two more apples than Ronnie. How many apples does Gayle have?*

Example 1 (1 minute): The Importance of Being Specific in Naming Variables

> **Example 1: The Importance of Being Specific in Naming Variables**
>
> When naming variables in expressions, it is important to be very clear about what they represent. The units of measure must be included if something is measured.

Exercises 1–2 (5 minutes)

Ask students to read the variables listed in the table and correct them for specificity.

> **Exercises 1–2**
>
> 1. Read the variable in the table, and improve the description given, making it more specific.
>
>   *Answers may vary because students may choose a different unit.*
>
> | Variable | Incomplete Description | Complete Description with Units |
> |---|---|---|
> | Joshua's speed (J) | Let J represent Joshua's speed. | *Let J represent Joshua's speed in meters per second.* |
> | Rufus's height (R) | Let R represent Rufus's height. | *Let R represent Rufus's height in centimeters.* |
> | Milk sold (M) | Let M represent the amount of milk sold. | *Let M represent the amount of milk sold in gallons.* |
> | Colleen's time in the 40-meter hurdles (C) | Let C represent Colleen's time. | *Let C represent Colleen's time in seconds.* |
> | Sean's age (S) | Let S represent Sean's age. | *Let S represent Sean's age in years.* |

- Again, when naming variables in expressions, it is important to be very clear about what they represent. When a variable represents a quantity of items, this too must be specified.

Review the concept of speed from the above table. Recall from Module 1 that speed is a rate. Emphasize that there are two different units needed to express speed (meters per second in the above example).

> 2. Read each variable in the table, and improve the description given, making it more specific.
>
> | Variable | Incomplete Description | Complete Description with Units |
> |---|---|---|
> | Karolyn's CDs (K) | Let K represent Karolyn's CDs. | Let K represent the number of CDs Karolyn has. |
> | Joshua's merit badges (J) | Let J represent Joshua's merit badges. | *Let J represent the number of merit badges Joshua has earned.* |
> | Rufus's trading cards (R) | Let R representRufus's trading cards. | *Let R represent the number of trading cards in Rufus's collection.* |
> | Milk money (M) | Let M represent the amount of milk money. | *Let M represent the amount of milk money collected in dollars.* |

Example 2 (17 minutes): Writing and Evaluating Addition and Subtraction Expressions

- Read the following story descriptions, and write an addition or subtraction expression for each one in the table.

Example 2: Writing and Evaluating Addition and Subtraction Expressions

Read each story problem. Identify the unknown quantity, and write the addition or subtraction expression that is described. Finally, evaluate your expression using the information given in column four.

Story Problem	Description with Units	Expression	Evaluate the Expression If:	Show Your Work and Evaluate
Gregg has two more dollars than his brother Jeff. Write an expression for the amount of money Gregg has.	Let j represent Jeff's money in dollars.	$j + 2$	Jeff has $12.	$j + 2$ $12 + 2$ 14 Gregg has $14.
Gregg has two more dollars than his brother Jeff. Write an expression for the amount of money Jeff has.	Let g represent Gregg's money in dollars.	$g - 2$	Gregg has $14.	$g - 2$ $14 - 2$ 12 Jeff has $12.
Abby read 8 more books than Kristen in the first marking period. Write an expression for the number of books Abby read.	*Let k represent the number of books Kristen read in the first marking period.*	$k + 8$	Kristen read 9 books in the first marking period.	$k + 8$ $9 + 8$ 17 *Abby read 17 books in the first marking period.*
Abby read 6 more books than Kristen in the second marking period. Write an expression for the number of books Kristen read.	*Let a represent the number of books Abby read in the second marking period.*	$a - 6$	Abby read 20 books in the second marking period.	$a - 6$ $20 - 6$ 14 *Kristen read 14 books in the second marking period.*
Daryl has been teaching for one year longer than Julie. Write an expression for the number of years that Daryl has been teaching.	*Let j represent the number of years Julie has been teaching.*	$j + 1$	Julie has been teaching for 28 years.	$j + 1$ $28 + 1$ 29 *Daryl has been teaching for 29 years.*
Ian scored 4 fewer goals than Julia in the first half of the season. Write an expression for the number of goals Ian scored.	*Let j represent the number of goals scored by Julia.*	$j - 4$	Julia scored 13 goals.	$j - 4$ $13 - 4$ 9 *Ian scored 9 goals in the first half of the season.*
Ian scored 3 fewer goals than Julia in the second half of the season. Write an expression for the number of goals Julia scored.	*Let i represent the number of goals scored by Ian.*	$i + 3$	Ian scored 8 goals.	$i + 3$ $8 + 3$ 11 *Julia scored 11 goals in the second half of the season.*
Johann visited Niagara Falls 3 times fewer than Arthur. Write an expression for the number of times Johann visited Niagara Falls.	*Let f represent the number of times Arthur visited Niagara Falls.*	$f - 3$	Arthur visited Niagara Falls 5 times.	$f - 3$ $5 - 3$ 2 *Johann visited Niagara Falls twice.*

Closing (5 minutes)

- Why is it important to describe the variable in an expression?

    □ *The biggest reason to define the variable is to know what the expression represents.*

    □ *If something is measured, include units. If something is counted, include that it is a number of items.*

- How do you determine if an expression will be an addition expression or a subtraction expression?

    □ *In the first problem in the table on the previous page, if we define x as the amount of money that Jeff has, then we would write an expression for the amount of money that Gregg has as $x + 2$. However, if we define the variable to be the amount of money that Gregg has, then we would write an expression to represent the amount of money that Jeff has as $x - 2$. Since the story problem represents a relationship between two quantities, both expressions are equally relevant.*

Exit Ticket (5 minutes)

Name _____   Date _____

Lesson 18: Writing and Evaluating Expressions—Addition and Subtraction

Exit Ticket

Kathleen lost a tooth today. Now she has lost 4 more than her sister Cara lost.

1. Write an expression to represent the number of teeth Cara has lost. Let K represent the number of teeth Kathleen lost.

    Expression:

2. Write an expression to represent the number of teeth Kathleen has lost. Let C represent the number of teeth Cara lost.

    Expression:

3. If Cara lost 3 teeth, how many teeth has Kathleen lost?

©2015 Great Minds. eureka-math.org
G6-M4-TE-B4-1.3.1-01.2016

Exit Ticket Sample Solutions

Kathleen lost a tooth today. Now she has lost 4 more than her sister Cara lost.

1. Write an expression to represent the number of teeth Cara has lost. Let K represent the number of teeth Kathleen lost.

    Expression: $K - 4$

2. Write an expression to represent the number of teeth Kathleen lost. Let C represent the number of teeth Cara lost.

    Expression: $C + 4$

3. If Cara lost 3 teeth, how many teeth has Kathleen lost?

    $C + 4;\ 3 + 4;$ *Kathleen has lost 7 teeth.*

Problem Set Sample Solutions

1. Read each story problem. Identify the unknown quantity, and write the addition or subtraction expression that is described. Finally, evaluate your expression using the information given in column four.

    Sample answers are shown. An additional expression can be written for each.

Story Problem	Description with Units	Expression	Evaluate the Expression If:	Show Your Work and Evaluate
Sammy has two more baseballs than his brother Ethan.	Let e represent the number of balls Ethan has.	$e + 2$	Ethan has 7 baseballs.	$e + 2$ $7 + 2$ 9 Sammy has 9 baseballs.
Ella wrote 8 more stories than Anna in the fifth grade.	*Let s represent the number of stories Anna wrote in the fifth grade.*	$s + 8$	Anna wrote 10 stories in the fifth grade.	$s + 8$ $10 + 8$ 18 *Ella wrote 18 stories in the fifth grade.*
Lisa has been dancing for 3 more years than Danika.	*Let y represent the number of years Danika has been dancing.*	$y + 3$	Danika has been dancing for 6 years.	$y + 3$ $6 + 3$ 9 *Lisa has been dancing for 9 years.*
The New York Rangers scored 2 fewer goals than the Buffalo Sabres last night.	*Let g represent the number of goals scored by the Rangers.*	$g + 2$	The Rangers scored 3 goals last night.	$g + 2$ $3 + 2$ 5 *The Buffalo Sabres scored 5 goals last night.*
George has gone camping 3 times fewer than Dave.	*Let c represent the number of times George has gone camping.*	$c + 3$	George has gone camping 8 times.	$c + 3$ $8 + 3$ 11 *Dave has gone camping 11 times.*

2. If George went camping 15 times, how could you figure out how many times Dave went camping?

    Adding 3 to George's camping trip total (15) would yield an answer of 18 trips for Dave.

Lesson 19: Substituting to Evaluate Addition and Subtraction Expressions

Student Outcomes

- Students develop expressions involving addition and subtraction from real-world problems.
- Students evaluate these expressions for given values.

Lesson Notes

In this lesson, students begin by filling in data tables to help them organize data and see patterns; then, they move to drawing their own tables. Encourage students to label the columns completely.

Classwork

Fluency Exercise (5 minutes): Subtraction of Decimals

Sprint: Refer to the Sprints and Sprint Delivery Script sections in the Module Overview for directions on how to administer a Sprint.

Opening Exercise (3 minutes)

Opening Exercise

My older sister is exactly two years older than I am. Sharing a birthday is both fun and annoying. Every year on our birthday, we have a party, which is fun, but she always brags that she is two years older than I am, which is annoying. Shown below is a table of our ages, starting when I was born:

My Age (in years)	My Sister's Age (in years)
0	2
1	3
2	4
3	5
4	6

Scaffolding:
Some students benefit from having blank tables prepared ahead of time.

Make sure students understand the context of the story problem. It should be clear that the day I was born was my sister's second birthday. My first birthday was her third birthday; my second birthday was her fourth birthday; and so on.

Discussion (5 minutes)

- Today in class, we will use data tables. They help us organize data and see patterns. We can use variables to make generalizations about the patterns we see.

---

a. **Looking at the table, what patterns do you see? Tell a partner.**

   My sister's age is always two years more than my age.

b. **On the day I turned 8 years old, how old was my sister?**

   10 years old

c. **How do you know?**

   Since my sister's age is always two years more than my age, we just add 2 to my age. $8 + 2 = 10$

d. **On the day I turned 16 years old, how old was my sister?**

   18 years old

e. **How do you know?**

   Since my sister's age is always two years more than my age, we just add 2 to my age. $16 + 2 = 18$

f. **Do we need to extend the table to calculate these answers?**

   No; the pattern is to add 2 to your age to calculate your sister's age.

Example 1 (5 minutes)

Example 1

My Age (in years)	My Sister's Age (in years)
0	2
1	3
2	4
3	5
4	6
Y	$Y + 2$

Scaffolding:

A number line in the classroom can provide an additional reference for students. A cardboard sheet with two windows cut to reveal Y and $Y + 2$ works with Examples 1 and 2.

a. **What if you don't know how old I am? Let's use a variable for my age. Let Y = my age in years. Can you develop an expression to describe how old my sister is?**

   Your sister is $Y + 2$ years old.

MP.6

b. **Please add that to the last row of the table.**

- My age is Y years. My sister is $Y + 2$ years old. So, no matter what my age is (or was), my sister's age in years will always be two years greater than mine.

Example 2 (5 minutes)

Example 2

My Age (in years)	My Sister's Age (in years)
0	2
1	3
2	4
3	5
4	6
$G - 2$	G

a.   How old was I when my sister was 6 years old?

4 years old

b.   How old was I when my sister was 15 years old?

13 years old

c.   How do you know?

My age is always 2 years less than my sister's age.

d.   Look at the table in Example 2. If you know my sister's age, can you determine my age?

We can subtract two from your sister's age, and that will equal your age.

e.   If we use the variable G for my sister's age in years, what expression would describe my age in years?

$G - 2$

f.   Fill in the last row of the table with the expressions.

My age is $G - 2$ years. My sister is G years old.

g.   With a partner, calculate how old I was when my sister was 22, 23, and 24 years old.

You were 20, 21, and 22 years old, respectively.

Exercises (15 minutes)

Exercises

1. Noah and Carter are collecting box tops for their school. They each bring in 1 box top per day starting on the first day of school. However, Carter had a head start because his aunt sent him 15 box tops before school began. Noah's grandma saved 10 box tops, and Noah added those on his first day.

    a. Fill in the missing values that indicate the total number of box tops each boy brought to school.

School Day	Number of Box Tops Noah Has	Number of Box Tops Carter Has
1	11	16
2	12	17
3	13	18
4	14	19
5	15	20

    b. If we let D be the number of days since the new school year began, on day D of school, how many box tops will Noah have brought to school?

    $D + 10$ box tops

    c. On day D of school, how many box tops will Carter have brought to school?

    $D + 15$ box tops

    d. On day 10 of school, how many box tops will Noah have brought to school?

    $10 + 10 = 20$; On day 10, Noah would have brought in 20 box tops.

    e. On day 10 of school, how many box tops will Carter have brought to school?

    $10 + 15 = 25$; On day 10, Carter would have brought in 25 box tops.

2. Each week the Primary School recycles 200 pounds of paper. The Intermediate School also recycles the same amount but had another 300 pounds left over from summer school. The Intermediate School custodian added this extra 300 pounds to the first recycle week.

    a. Number the weeks, and record the amount of paper recycled by both schools.

Week	Total Amount of Paper Recycled by the Primary School This School Year in Pounds	Total Amount of Paper Recycled by the Intermediate School This School Year in Pounds
1	200	500
2	400	700
3	600	900
4	800	1,100
5	1,000	1,300

    b. If this trend continues, what will be the total amount collected for each school on Week 10?

    The Primary School will have collected $2,000$ pounds. The Intermediate School will have collected $2,300$ pounds.

3. Shelly and Kristen share a birthday, but Shelly is 5 years older.

    a. Make a table showing their ages every year, beginning when Kristen was born.

Kristen's Age (in years)	Shelly's Age (in years)
0	5
1	6
2	7
3	8

    b. If Kristen is 16 years old, how old is Shelly?

       If Kristen is 16 years old, Shelly is 21 years old.

    c. If Kristen is K years old, how old is Shelly?

       If Kristen is K years old, Shelly is $K + 5$ years old.

    d. If Shelly is S years old, how old is Kristen?

       If Shelly is S years old, Kristen is $S - 5$ years old.

Closing (2 minutes)

- Why were we able to write these expressions?
  - *There was a relationship between the two quantities that we could identify.*
- What is important to remember about labeling columns in a table?
  - *The label should be complete, with units, so the reader understands precisely what is meant.*
- How are addition and subtraction expressions related to one another?
  - *They are inverse operations. One undoes the other.*

Exit Ticket (5 minutes)

Name _____   Date _____

Lesson 19: Substituting to Evaluate Addition and Subtraction Expressions

Exit Ticket

Jenna and Allie work together at a piano factory. They both were hired on January 3, but Jenna was hired in 2005, and Allie was hired in 2009.

a. Fill in the table below to summarize the two workers' experience totals.

Year	Allie's Years of Experience	Jenna's Years of Experience
2010		
2011		
2012		
2013		
2014		

b. If both workers continue working at the piano factory, when Allie has A years of experience on the job, how many years of experience will Jenna have on the job?

c. If both workers continue working at the piano factory, when Allie has 20 years of experience on the job, how many years of experience will Jenna have on the job?

Exit Ticket Sample Solutions

Jenna and Allie work together at a piano factory. They both were hired on January 3, but Jenna was hired in 2005, and Allie was hired in 2009.

a. Fill in the table below to summarize the two workers' experience totals.

Year	Allie's Years of Experience	Jenna's Years of Experience
2010	1	5
2011	2	6
2012	3	7
2013	4	8
2014	5	9

b. If both workers continue working at the piano factory, when Allie has A years of experience on the job, how many years of experience will Jenna have on the job?

Jenna will have been on the job for $A + 4$ years.

c. If both workers continue working at the piano factory, when Allie has 20 years of experience on the job, how many years of experience will Jenna have on the job?

$20 + 4 = 24$

Jenna will have been on the job for 24 years.

Problem Set Sample Solutions

1. Suellen and Tara are in sixth grade, and both take dance lessons at Twinkle Toes Dance Studio. This is Suellen's first year, while this is Tara's fifth year of dance lessons. Both girls plan to continue taking lessons throughout high school.

a. Complete the table showing the number of years the girls will have danced at the studio.

Grade	Suellen's Years of Experience Dancing	Tara's Years of Experience Dancing
Sixth	1	5
Seventh	2	6
Eighth	3	7
Ninth	4	8
Tenth	5	9
Eleventh	6	10
Twelfth	7	11

b. If Suellen has been taking dance lessons for Y years, how many years has Tara been taking lessons?

Tara has been taking dance lessons for $Y + 4$ years.

©2015 Great Minds. eureka-math.org
G6-M4-TE-B4-1.3.1-01.2016

2. Daejoy and Damian collect fossils. Before they went on a fossil-hunting trip, Daejoy had 25 fossils in her collection, and Damian had 16 fossils in his collection. On a 10-day fossil-hunting trip, they each collected 2 new fossils each day.

   a. Make a table showing how many fossils each person had in their collection at the end of each day.

Day	Number of Fossils in Daejoy's Collection	Number of Fossils in Damian's Collection
1	27	18
2	29	20
3	31	22
4	33	24
5	35	26
6	37	28
7	39	30
8	41	32
9	43	34
10	45	36

   b. If this pattern of fossil finding continues, how many fossils does Damian have when Daejoy has F fossils?

When Daejoy has F fossils, Damian has $F - 9$ fossils.

   c. If this pattern of fossil finding continues, how many fossils does Damian have when Daejoy has 55 fossils?

$55 - 9 = 46$

When Daejoy has 55 fossils, Damian has 46 fossils.

3. A train consists of three types of cars: box cars, an engine, and a caboose. The relationship among the types of cars is demonstrated in the table below.

Number of Box Cars	Number of Cars in the Train
0	2
1	3
2	4
10	12
100	102

   a. Tom wrote an expression for the relationship depicted in the table as $B + 2$. Theresa wrote an expression for the same relationship as $C - 2$. Is it possible to have two different expressions to represent one relationship? Explain.

Both expressions can represent the same relationship, depending on the point of view. The expression $B + 2$ represents the number of box cars plus an engine and a caboose. The expression $C - 2$ represents the whole car length of the train, less the engine and caboose.

   b. What do you think the variable in each student's expression represents? How would you define them?

The variable C would represent the total cars in the train. The variable B would represent the number of box cars.

©2015 Great Minds. eureka-math.org
G6-M4-TE-B4-1.3.1-01.2016

4. David was 3 when Marieka was born. Complete the table.

Marieka's Age in Years	David's Age in Years
5	8
6	9
7	10
8	11
10	13
17	20
32	35
M	$M + 3$
$D - 3$	D

5. Caitlin and Michael are playing a card game. In the first round, Caitlin scored 200 points, and Michael scored 175 points. In each of the next few rounds, they each scored 50 points. Their score sheet is below.

Caitlin's Points	Michael's Points
200	175
250	225
300	275
350	325

a. If this trend continues, how many points will Michael have when Caitlin has 600 points?

$600 - 25 = 575$

Michael will have 575 points.

b. If this trend continues, how many points will Michael have when Caitlin has C points?

Michael will have $C - 25$ points.

c. If this trend continues, how many points will Caitlin have when Michael has 975 points?

$975 + 25 = 1000$

Caitlin will have $1,000$ points.

d. If this trend continues, how many points will Caitlin have when Michael has M points?

Caitlin will have $M + 25$ points.

6. The high school marching band has 15 drummers this year. The band director insists that there are to be 5 more trumpet players than drummers at all times.

a. How many trumpet players are in the marching band this year?

$15 + 5 = 20$. *There are 20 trumpet players this year.*

b. Write an expression that describes the relationship of the number of trumpet players (T) and the number of drummers (D).

$T = D + 5 \text{ or } D = T - 5$

c. If there are only 14 trumpet players interested in joining the marching band next year, how many drummers will the band director want in the band?

$14 - 5 = 9$.

The band director will want 9 drummers.

Number Correct: _____

Subtraction of Decimals—Round 1

Directions: Evaluate each expression.

1.	$55 - 50$	
2.	$55 - 5$	
3.	$5.5 - 5$	
4.	$5.5 - 0.5$	
5.	$88 - 80$	
6.	$88 - 8$	
7.	$8.8 - 8$	
8.	$8.8 - 0.8$	
9.	$33 - 30$	
10.	$33 - 3$	
11.	$3.3 - 3$	
12.	$1 - 0.3$	
13.	$1 - 0.03$	
14.	$1 - 0.003$	
15.	$0.1 - 0.03$	
16.	$4 - 0.8$	
17.	$4 - 0.08$	
18.	$4 - 0.008$	
19.	$0.4 - 0.08$	
20.	$9 - 0.4$	
21.	$9 - 0.04$	
22.	$9 - 0.004$	

23.	$9.9 - 5$	
24.	$9.9 - 0.5$	
25.	$0.99 - 0.5$	
26.	$0.99 - 0.05$	
27.	$4.7 - 2$	
28.	$4.7 - 0.2$	
29.	$0.47 - 0.2$	
30.	$0.47 - 0.02$	
31.	$8.4 - 1$	
32.	$8.4 - 0.1$	
33.	$0.84 - 0.1$	
34.	$7.2 - 5$	
35.	$7.2 - 0.5$	
36.	$0.72 - 0.5$	
37.	$0.72 - 0.05$	
38.	$8.6 - 7$	
39.	$8.6 - 0.7$	
40.	$0.86 - 0.7$	
41.	$0.86 - 0.07$	
42.	$5.1 - 4$	
43.	$5.1 - 0.4$	
44.	$0.51 - 0.4$	

Subtraction of Decimals—Round 1 [KEY]

Directions: Evaluate each expression.

1.	$55 - 50$	**5**	23.	$9.9 - 5$	**4.9**
2.	$55 - 5$	**50**	24.	$9.9 - 0.5$	**9.4**
3.	$5.5 - 5$	**0.5**	25.	$0.99 - 0.5$	**0.49**
4.	$5.5 - 0.5$	**5**	26.	$0.99 - 0.05$	**0.94**
5.	$88 - 80$	**8**	27.	$4.7 - 2$	**2.7**
6.	$88 - 8$	**80**	28.	$4.7 - 0.2$	**4.5**
7.	$8.8 - 8$	**0.8**	29.	$0.47 - 0.2$	**0.27**
8.	$8.8 - 0.8$	**8**	30.	$0.47 - 0.02$	**0.45**
9.	$33 - 30$	**3**	31.	$8.4 - 1$	**7.4**
10.	$33 - 3$	**30**	32.	$8.4 - 0.1$	**8.3**
11.	$3.3 - 3$	**0.3**	33.	$0.84 - 0.1$	**0.74**
12.	$1 - 0.3$	**0.7**	34.	$7.2 - 5$	**2.2**
13.	$1 - 0.03$	**0.97**	35.	$7.2 - 0.5$	**6.7**
14.	$1 - 0.003$	**0.997**	36.	$0.72 - 0.5$	**0.22**
15.	$0.1 - 0.03$	**0.07**	37.	$0.72 - 0.05$	**0.67**
16.	$4 - 0.8$	**3.2**	38.	$8.6 - 7$	**1.6**
17.	$4 - 0.08$	**3.92**	39.	$8.6 - 0.7$	**7.9**
18.	$4 - 0.008$	**3.992**	40.	$0.86 - 0.7$	**0.16**
19.	$0.4 - 0.08$	**0.32**	41.	$0.86 - 0.07$	**0.79**
20.	$9 - 0.4$	**8.6**	42.	$5.1 - 4$	**1.1**
21.	$9 - 0.04$	**8.96**	43.	$5.1 - 0.4$	**4.7**
22.	$9 - 0.004$	**8.996**	44.	$0.51 - 0.4$	**0.11**

EUREKA
MATH

Number Correct: _____

Improvement: _____

Subtraction of Decimals—Round 2

Directions: Evaluate each expression.

1.	$66 - 60$	
2.	$66 - 6$	
3.	$6.6 - 6$	
4.	$6.6 - 0.6$	
5.	$99 - 90$	
6.	$99 - 9$	
7.	$9.9 - 9$	
8.	$9.9 - 0.9$	
9.	$22 - 20$	
10.	$22 - 2$	
11.	$2.2 - 2$	
12.	$3 - 0.4$	
13.	$3 - 0.04$	
14.	$3 - 0.004$	
15.	$0.3 - 0.04$	
16.	$8 - 0.2$	
17.	$8 - 0.02$	
18.	$8 - 0.002$	
19.	$0.8 - 0.02$	
20.	$5 - 0.1$	
21.	$5 - 0.01$	
22.	$5 - 0.001$	

23.	$6.8 - 4$	
24.	$6.8 - 0.4$	
25.	$0.68 - 0.4$	
26.	$0.68 - 0.04$	
27.	$7.3 - 1$	
28.	$7.3 - 0.1$	
29.	$0.73 - 0.1$	
30.	$0.73 - 0.01$	
31.	$9.5 - 2$	
32.	$9.5 - 0.2$	
33.	$0.95 - 0.2$	
34.	$8.3 - 5$	
35.	$8.3 - 0.5$	
36.	$0.83 - 0.5$	
37.	$0.83 - 0.05$	
38.	$7.2 - 4$	
39.	$7.2 - 0.4$	
40.	$0.72 - 0.4$	
41.	$0.72 - 0.04$	
42.	$9.3 - 7$	
43.	$9.3 - 0.7$	
44.	$0.93 - 0.7$	

Subtraction of Decimals—Round 2 [KEY]

Directions: Evaluate each expression.

1.	$66 - 60$	**6**	23.	$6.8 - 4$	**2.8**
2.	$66 - 6$	**60**	24.	$6.8 - 0.4$	**6.4**
3.	$6.6 - 6$	**0.6**	25.	$0.68 - 0.4$	**0.28**
4.	$6.6 - 0.6$	**6**	26.	$0.68 - 0.04$	**0.64**
5.	$99 - 90$	**9**	27.	$7.3 - 1$	**6.3**
6.	$99 - 9$	**90**	28.	$7.3 - 0.1$	**7.2**
7.	$9.9 - 9$	**0.9**	29.	$0.73 - 0.1$	**0.63**
8.	$9.9 - 0.9$	**9**	30.	$0.73 - 0.01$	**0.72**
9.	$22 - 20$	**2**	31.	$9.5 - 2$	**7.5**
10.	$22 - 2$	**20**	32.	$9.5 - 0.2$	**9.3**
11.	$2.2 - 2$	**0.2**	33.	$0.95 - 0.2$	**0.75**
12.	$3 - 0.4$	**2.6**	34.	$8.3 - 5$	**3.3**
13.	$3 - 0.04$	**2.96**	35.	$8.3 - 0.5$	**7.8**
14.	$3 - 0.004$	**2.996**	36.	$0.83 - 0.5$	**0.33**
15.	$0.3 - 0.04$	**0.26**	37.	$0.83 - 0.05$	**0.78**
16.	$8 - 0.2$	**7.8**	38.	$7.2 - 4$	**3.2**
17.	$8 - 0.02$	**7.98**	39.	$7.2 - 0.4$	**6.8**
18.	$8 - 0.002$	**7.998**	40.	$0.72 - 0.4$	**0.32**
19.	$0.8 - 0.02$	**0.78**	41.	$0.72 - 0.04$	**0.68**
20.	$5 - 0.1$	**4.9**	42.	$9.3 - 7$	**2.3**
21.	$5 - 0.01$	**4.99**	43.	$9.3 - 0.7$	**8.6**
22.	$5 - 0.001$	**4.999**	44.	$0.93 - 0.7$	**0.23**

EUREKA
MATH™

Lesson 20: Writing and Evaluating Expressions— Multiplication and Division

Student Outcomes

- Students develop expressions involving multiplication and division from real-world problems.
- Students evaluate these expressions for given values.

Lesson Notes

This lesson builds on Lessons 18 and 19, extending the concepts using multiplication and division expressions.

Classwork

Opening (3 minutes)

Take time to make sure the answers to the Problem Set from the previous lesson are clear. The labels on the tables should be complete.

Discussion (3 minutes)

- In the previous lessons, we created expressions that used addition and subtraction to describe the relationship between two quantities. How did using tables help your understanding?
    - *Answers will vary. Patterns were easy to see. Looking down the columns revealed a number pattern. Looking across the rows revealed a constant difference between columns.*
- In this lesson, we are going to develop expressions involving multiplication and division, much like the last lesson. We also evaluate these expressions for given values.

Example 1 (10 minutes)

- The farmers' market is selling bags of apples. In every bag, there are 3 apples. If I buy one bag, how many apples will I have?
    - *Three*
- If I buy two bags, how many apples will I have?
    - *Since $2 \cdot 3 = 6$, you will have 6 apples.*
- If I buy three bags, how many apples will I have?
    - *Since $3 \cdot 3 = 9$, you will have 9 apples.*
- Fill in the table for a purchase of 4 bags of apples. Check your answer with a partner.

> **Scaffolding:**
> Having interlocking cubes ready in groups of three makes a concrete visual for students to see and hold for Example 1. Put these in clear plastic bags, if desired.

Example 1

The farmers' market is selling bags of apples. In every bag, there are 3 apples.

a. Complete the table.

Number of Bags	Total Number of Apples
1	3
2	6
3	9
4	12
B	$3B$

- What if I bought some other number of bags? If I told you how many bags, could you calculate the number of apples I would have altogether?

  □ *Yes. Multiply the number of bags by 3 to find the total number of apples.*

- What if I bought B bags of apples? Can you write an expression in the table that describes the total number of apples I have purchased?

  □ *$3B$ or $3(B)$ or $3 \cdot B$*

Take a moment to review the different notations used for multiplication. Students should be comfortable reading and writing the expressions in all three forms.

- What if the market had 25 bags of apples to sell? How many apples is that in all?

  □ *If $B = 25$, then $3B = 3 \cdot 25 = 75$. The market had 75 apples to sell.*

b. What if the market had 25 bags of apples to sell? How many apples is that in all?

  If $B = 25$, then $3B = 3 \cdot 25 = 75$. The market had 75 apples to sell.

c. If a truck arrived that had some number, a, more apples on it, then how many bags would the clerks use to bag up the apples?

  $a \div 3$ bags are needed. If there are 1 or 2 apples left over, an extra bag will be needed (although not full).

d. If a truck arrived that had 600 apples on it, how many bags would the clerks use to bag up the apples?

  $$600 \text{ apples} \div 3 \frac{\text{apples}}{\text{bag}} = 200 \text{ bags}$$

e. How is part (d) different from part (b)?

  Part (d) gives the number of apples and asks for the number of bags. Therefore, we needed to divide the number of apples by 3. Part (b) gives the number of bags and asks for the number of apples. Therefore, we needed to multiply the number of bags by 3.

Exercise 1 (5 minutes)

Students work on Exercise 1 independently.

Exercises 1–3

1. In New York State, there is a five-cent deposit on all carbonated beverage cans and bottles. When you return the empty can or bottle, you get the five cents back.

    a. Complete the table.

Number of Containers Returned	Refund in Dollars
1	0.05
2	0.10
3	0.15
4	0.20
10	0.50
50	2.50
100	5.00
C	$0.05C$

    b. If we let C represent the number of cans, what is the expression that shows how much money is returned?

    $0.05C$

    c. Use the expression to find out how much money Brett would receive if he returned 222 cans.

    If $C = 222$, then $0.05C = 0.05 \cdot 222 = 11.10$. Brett would receive $\$11.10$ if he returned 222 cans.

    d. If Gavin needs to earn $\$4.50$ for returning cans, how many cans does he need to collect and return?

    $4.50 \div 0.05 = 90$. Gavin needs to collect and return 90 cans.

    e. How is part (d) different from part (c)?

    Part (d) gives the amount of money and asks for the number of cans. Therefore, we needed to divide the amount of money by 0.05. Part (c) gives the number of cans and asks for the amount of money. Therefore, we needed to multiply the number of cans by 0.05.

Discuss the similarities and differences between Example 1 and Exercise 1. In both problems, the second quantity is a multiple of the first. Multiplication by the constant term is used to show the relationship between the quantities in the first column and the quantities in the second column. Division is used to show the relationship between the quantities in the second column and the quantities in the first column.

Exercise 2 (10 minutes)

Students work on Exercise 2 independently.

---

2.   The fare for a subway or a local bus ride is $2.50.

a.   Complete the table.

Number of Rides	Cost of Rides in Dollars
1	2.50
2	5.00
3	7.50
4	10.00
5	12.50
10	25.00
30	75.00
R	$2.50R$ or $2.5R$

b.   If we let R represent the number of rides, what is the expression that shows the cost of the rides?

$2.50R$ or $2.5R$

c.   Use the expression to find out how much money 60 rides would cost.

If $R = 60$, then $2.50R = 2.50 \cdot 60 = 150.00$. Sixty rides would cost $150.00.

d.   If a commuter spends $175.00 on subway or bus rides, how many trips did the commuter take?

$175.00 \div 2.50 = 70$. The commuter took 70 trips.

e.   How is part (d) different from part (c)?

Part (d) gives the amount of money and asks for the number of rides. Therefore, we needed to divide the amount of money by the cost of each ride ($2.50). Part (c) gives the number of rides and asks for the amount of money. Therefore, we needed to multiply the number of rides by $2.50.

Exercise 3 (10 minutes): Challenge Problem

> **Challenge Problem**
>
> 3.   A pendulum swings though a certain number of cycles in a given time. Owen made a pendulum that swings 12 times every 15 seconds.
>
> a.   Construct a table showing the number of cycles through which a pendulum swings. Include data for up to one minute. Use the last row for C cycles, and write an expression for the time it takes for the pendulum to make C cycles.
>
> | Number of Cycles | Time in Seconds |
> |:---:|:---:|
> | 12 | 15 |
> | 24 | 30 |
> | 36 | 45 |
> | 48 | 60 |
> | C | $\dfrac{15C}{12}$ |
>
> b.   Owen and his pendulum team set their pendulum in motion and counted 16 cycles. What was the elapsed time?
>
> $C = 16; \dfrac{15 \cdot 16}{12} = 20$. *The elapsed time is* 20 *seconds.*
>
> c.   Write an expression for the number of cycles a pendulum swings in S seconds.
>
> $\dfrac{12}{15} S$ *or* $\dfrac{4}{5} S$ *or* $0.8 \cdot S$
>
> d.   In a different experiment, Owen and his pendulum team counted the cycles of the pendulum for 35 seconds. How many cycles did they count?
>
> $S = 35; 0.8 \cdot 35 = 28$. *They counted* 28 *cycles.*

Closing (2 minutes)

- In Example 1, we looked at the relationship between the number of bags purchased at the farmers' market and the total number of apples purchased. We created two different expressions: $3B$ and $a \div 3$. What does each variable represent, and why did we multiply by 3 in the first expression and divide by 3 in the second?
  - *The variable B represented the number of bags. We had to multiply by 3 because we were given the number of bags, and there were 3 apples packaged in each bag. The variable a represented the number of apples. We divided by 3 because we were given the number of apples and need to determine the number of bags needed.*
- What would the expressions be if the farmers' market sold bags that contained 5 apples in a bag instead of 3?
  - *$5B$ and $a \div 5$, respectively*

Exit Ticket (3 minutes)

Name _____   Date _____

Lesson 20: Writing and Evaluating Expressions—Multiplication and Division

Exit Ticket

Anna charges $8.50 per hour to babysit. Complete the table, and answer the questions below.

Number of Hours	Amount Anna Charges in Dollars
1	
2	
5	
8	
H	

a.   Write an expression describing her earnings for working H hours.

b.   How much will she earn if she works for $3\frac{1}{2}$ hours?

c.   How long will it take Anna to earn $51.00?

EUREKA
MATH™

Exit Ticket Sample Solutions

1. Anna charges 8.50 per hour to babysit. Complete the table, and answer the questions below.

Number of Hours	Amount Anna Charges in Dollars
1	8.50
2	17.00
5	42.50
8	68
H	$8.50H$ or $8.5H$

    a. Write an expression describing her earnings for working H hours.

       $8.50H$ or $8.5H$

    b. How much will she earn if she works for $3\frac{1}{2}$ hours?

       If $H = 3.5$, then $8.5H = 8.5 \cdot 3.5 = 29.75$. She will earn $\$29.75$.

    c. How long will it take Anna to earn $\$51.00$?

       $51 \div 8.5 = 6$. It will take Anna 6 hours to earn $\$51.00$.

Problem Set Sample Solutions

1. A radio station plays 12 songs each hour. They never stop for commercials, news, weather, or traffic reports.

    a. Write an expression describing how many songs are played by the radio station in H hours.

       $12H$

    b. How many songs will be played in an entire day (24 hours)?

       $12 \cdot 24 = 288$. There will be 288 songs played.

    c. How long does it take the radio station to play 60 consecutive songs?

       $60 \text{ songs} \div 12\frac{\text{songs}}{\text{hour}} = 5 \text{ hours}$

2. A ski area has a high-speed lift that can move $2,400$ skiers to the top of the mountain each hour.

    a. Write an expression describing how many skiers can be lifted in H hours.

       $2,400H$

    b. How many skiers can be moved to the top of the mountain in 14 hours?

       $14 \cdot 2,400 = 33,600$. $33,600$ skiers can be moved.

©2015 Great Minds. eureka-math.org
G6-M4-TE-B4-1.3.1-01.2016

c. How long will it take to move $3,600$ skiers to the top of the mountain?

$3,600 \div 2,400 = 1.5.$ *It will take an hour and a half to move $3,600$ skiers to the top of the mountain.*

3. Polly writes a magazine column, for which she earns $35 per hour. Create a table of values that shows the relationship between the number of hours that Polly works, H, and the amount of money Polly earns in dollars, E.

Answers will vary. Sample answers are shown.

Hours Polly Works (H)	Polly's Earnings in Dollars (E)
1	35
2	70
3	105
4	140

a. If you know how many hours Polly works, can you determine how much money she earned? Write the corresponding expression.

Multiplying the number of hours that Polly works by her rate ($35 per hour) will calculate her pay. $35H$ is the expression for her pay in dollars.

b. Use your expression to determine how much Polly earned after working for $3\frac{1}{2}$ hours.

$35H = 35 \cdot 3.5 = 122.5.$ *Polly makes $122.50 for working $3\frac{1}{2}$ hours.*

c. If you know how much money Polly earned, can you determine how long she worked? Write the corresponding expression.

Dividing Polly's pay by 35 will calculate the number of hours she worked. $E \div 35$ is the expression for the number of hours she worked.

d. Use your expression to determine how long Polly worked if she earned $52.50.

$52.50 \div 35 = 1.5;$ *Polly worked an hour and a half for $52.50.*

4. Mitchell delivers newspapers after school, for which he earns $0.09 per paper. Create a table of values that shows the relationship between the number of papers that Mitchell delivers, P, and the amount of money Mitchell earns in dollars, E.

Answers will vary. Sample answers are shown.

Number of Papers Delivered (P)	Mitchell's Earnings in Dollars (E)
1	0.09
10	0.90
100	9.00
1,000	90.00

a. If you know how many papers Mitchell delivered, can you determine how much money he earned? Write the corresponding expression.

Multiplying the number of papers that Mitchell delivers by his rate ($0.09 per paper) will calculate his pay. $0.09P$ is the expression for his pay in dollars.

Lesson 20: Writing and Evaluating Expressions—Multiplication and Division

b. Use your expression to determine how much Mitchell earned by delivering 300 newspapers.

$0.09P = 0.09 \cdot 300 = 27$. *Mitchell earned* 27.00 *for delivering* 300 *newspapers.*

c. If you know how much money Mitchell earned, can you determine how many papers he delivered? Write the corresponding expression.

Dividing Mitchell's pay by 0.09 *will calculate the number of papers he delivered.* $E \div 0.09$ *is the expression for the number of papers he delivered.*

d. Use your expression to determine how many papers Mitchell delivered if he earned $58.50 last week.

$58.50 \div 0.09 = 650$; *therefore, Mitchell delivered* 650 *newspapers last week.*

5. Randy is an art dealer who sells reproductions of famous paintings. Copies of the *Mona Lisa* sell for $475.

a. Last year Randy sold $9,975 worth of *Mona Lisa* reproductions. How many did he sell?

$9,975 \div 475 = 21$. *He sold* 21 *copies of the painting.*

b. If Randy wants to increase his sales to at least $15,000 this year, how many copies will he need to sell (without changing the price per painting)?

$15,000 \div 475$ *is about* 31.6. *He will have to sell* 32 *paintings in order to increase his sales to at least* $15,000.

Lesson 21: Writing and Evaluating Expressions—Multiplication and Addition

Student Outcomes

- Students develop formulas involving multiplication and addition from real-world problems.
- Students evaluate these formulas for given values.

Lesson Notes

This lesson begins with students making a model of a real-world problem they most likely have already encountered: moving enough tables together so that a large group of people can sit together. After the problem is posed, students use square tiles to model the problem. Using this data and looking for patterns, they make generalizations about the expression that describes the problem.

It is necessary to prepare bags of five square tiles before class and to group students thoughtfully.

Classwork

Opening (2 minutes)

Move students into groups of two or three, and distribute the bags of tiles.

Mathematical Modeling Exercise (15 minutes)

- Today, we will model a problem that happens in restaurants every day: moving tables together so that everyone in a group can sit together. Use the square tiles to represent square tables. One person can sit along each edge of a table side, no crowding. Our first goal is to find how many people can sit at tables made of various numbers of square tables pushed together end to end.

- How many chairs can fit around one square table? What is the perimeter of the square if the edge length is one yard? Record the results in your table.
  - *4; 4 yards*

- If two square tables are pushed together to form a longer rectangular table, how many chairs will fit around the new table? What is the perimeter of the rectangle? Record the results in your table.
  - *6; 6 yards*

Make sure that each student can connect the square model on the desk to the picture on the classwork sheet.

- If there are twice as many square tables in the new rectangular table, why can't twice as many chairs fit around it?
  - *No chairs will fit right where the tables come together.*

- Make a record of the number of chairs that will fit around longer rectangular tables when 3, 4, and 5 square tables are pushed together to form long rectangular tables.

Mathematical Modeling Exercise

The Italian Villa Restaurant has square tables that the servers can push together to accommodate the customers. Only one chair fits along the side of the square table. Make a model of each situation to determine how many seats will fit around various rectangular tables.

Number of Square Tables	Number of Seats at the Table
1	4
2	6
3	8
4	10
5	12
50	102
200	402
T	$2T + 2$ or $2(T + 1)$

Are there any other ways to think about solutions to this problem?

Regardless of the number of tables, there is one chair on each end, and each table has two chairs opposite one another.

It is impractical to make a model of pushing 50 tables together to make a long rectangle. If we did have a rectangle that long, how many chairs would fit on the long sides of the table?

50 on each side, for a total of 100

How many chairs fit on the ends of the long table?

2 chairs, one on each end

How many chairs fit in all? Record it on your table.

102 chairs in all

Work with your group to determine how many chairs would fit around a very long rectangular table if 200 square tables were pushed together.

200 chairs on each side, totaling 400, plus one on each end; grand total 402

If we let T represent the number of square tables that make one long rectangular table, what is the expression for the number of chairs that will fit around it?

$2T + 2$

Example (13 minutes)

> **Example**
>
> Look at Example 1 with your group. Determine the cost for various numbers of pizzas, and also determine the expression that describes the cost of having P pizzas delivered.
>
> a. Pizza Queen has a special offer on lunch pizzas: $\$4.00$ each. They charge $\$2.00$ to deliver, regardless of how many pizzas are ordered. Determine the cost for various numbers of pizzas, and also determine the expression that describes the cost of having P pizzas delivered.
>
> | Number of Pizzas Delivered | Total Cost in Dollars |
> | :---: | :---: |
> | 1 | 6 |
> | 2 | 10 |
> | 3 | 14 |
> | 4 | 18 |
> | 10 | 42 |
> | 50 | 202 |
> | P | $4P + 2$ |

Allow the groups to discover patterns and share them.

> What mathematical operations did you need to perform to find the total cost?
>
> *Multiplication and addition. We multiplied the number of pizzas by $\$4$ and then added the $\$2$ delivery fee.*
>
> Suppose our principal wanted to buy a pizza for everyone in our class. Determine how much this would cost.
>
> *Answers will vary depending on the number of students in your class.*
>
> b. If the booster club had $\$400$ to spend on pizza, what is the greatest number of pizzas they could order?

Students can use the "guess and check" method for answering this question. A scaffold question might be, "Could they order 100 pizzas at this price?"

> *The greatest number of pizzas they could order would be 99. The pizzas themselves would cost $99 \times \$4 = \$396$, and then add $\$2.00$ for delivery. The total bill is $\$398$.*

c. If the pizza price was raised to $5.00 and the delivery price was raised to $3.00, create a table that shows the total cost (pizza plus delivery) of 1, 2, 3, 4, and 5 pizzas. Include the expression that describes the new cost of ordering P pizzas.

Number of Pizzas Delivered	Total Cost in Dollars
1	8
2	13
3	18
4	23
5	28
P	$5P + 3$

Closing (8 minutes)

- Some mathematical expressions use both multiplication and addition. With your partner, make up a new example of a problem that uses both multiplication and addition.

Allow a short time for groups to make up a situation. Share these as a group. Ensure that there is both a coefficient and a constant in each problem. Naming these terms is not important for this lesson.

Exit Ticket (7 minutes)

©2015 Great Minds. eureka-math.org
G6-M4-TE-B4-1.3.1-01.2016

Name _____   Date _____

Lesson 21: Writing and Evaluating Expressions—Multiplication and Addition

Exit Ticket

Krystal Klear Cell Phone Company charges $5.00 per month for service. The company also charges $0.10 for each text message sent.

a. Complete the table below to calculate the monthly charges for various numbers of text messages sent.

Number of Text Messages Sent (T)	Total Monthly Bill in Dollars
0	
10	
20	
30	
T	

b. If Suzannah's budget limit is $10 per month, how many text messages can she send in one month?

Exit Ticket Sample Solutions

Krystal Klear Cell Phone Company charges $5.00 per month for service. The company also charges $0.10 for each text message sent.

    a.   Complete the table below to calculate the monthly charges for various numbers of text messages sent.

Number of Text Messages Sent (T)	Total Monthly Bill in Dollars
0	5
10	6
20	7
30	8
T	$0.1T + 5$

    b.   If Suzannah's budget limit is $10 per month, how many text messages can she send in one month?

        Suzannah can send 50 text messages in one month for $10.

Problem Set Sample Solutions

1.   Compact discs (CDs) cost $12 each at the Music Emporium. The company charges $4.50 for shipping and handling, regardless of how many compact discs are purchased.

    a.   Create a table of values that shows the relationship between the number of compact discs that Mickey buys, D, and the amount of money Mickey spends, C, in dollars.

Number of CDs Mickey Buys (D)	Total Cost in Dollars (C)
1	$16.50
2	$28.50
3	$40.50

    b.   If you know how many CDs Mickey orders, can you determine how much money he spends? Write the corresponding expression.

        $12D + 4.5$

    c.   Use your expression to determine how much Mickey spent buying 8 CDs.

        $8(12) + 4.50 = 100.50.$ *Mickey spent $100.50.*

2.   Mr. Gee's class orders paperback books from a book club. The books cost $2.95 each. Shipping charges are set at $4.00, regardless of the number of books purchased.

    a.   Create a table of values that shows the relationship between the number of books that Mr. Gee's class buys, B, and the amount of money they spend, C, in dollars.

Number of Books Ordered(B)	Amount of Money Spent in Dollars (C)
1	6.95
2	9.90
3	12.85

©2015 Great Minds. eureka-math.org
G6-M4-TE-B4-1.3.1-01.2016

b. If you know how many books Mr. Gee's class orders, can you determine how much money they spend? Write the corresponding expression.

$2.95B + 4$

c. Use your expression to determine how much Mr. Gee's class spent buying 24 books.

$24(2.95) + 4 = 74.$ *Mr. Gee's class spent* $74.80.

3. Sarah is saving money to take a trip to Oregon. She received $450 in graduation gifts and saves $120 per week working.

a. Write an expression that shows how much money Sarah has after working W weeks.

$450 + 120W$

b. Create a table that shows the relationship between the amount of money Sarah has (M) and the number of weeks she works (W).

Amount of Money Sarah Has (M)	Number of Weeks Worked (W)
570	1
690	2
810	3
930	4
1,050	5
1,170	6
1,290	7
1,410	8

c. The trip will cost $1,200. How many weeks will Sarah have to work to earn enough for the trip?

Sarah will have to work 7 weeks to earn enough for the trip.

4. Mr. Gee's language arts class keeps track of how many words per minute are read aloud by each of the students. They collect this oral reading fluency data each month. Below is the data they collected for one student in the first four months of school.

a. Assume this increase in oral reading fluency continues throughout the rest of the school year. Complete the table to project the reading rate for this student for the rest of the year.

Month	Number of Words Read Aloud in One Minute
September	126
October	131
November	136
December	141
January	146
February	151
March	156
April	161
May	166
June	171

©2015 Great Minds. eureka-math.org
G6-M4-TE-B4-1.3.1-01.2016

b. If this increase in oral reading fluency continues throughout the rest of the school year, when would this student achieve the goal of reading 165 words per minute?

The student will meet the goal in May.

c. The expression for this student's oral reading fluency is $121 + 5m$, where m represents the number of months during the school year. Use this expression to determine how many words per minute the student would read after 12 months of instruction.

The student would read 181 words per minute: $121 + 5 \times 12$.

5. When corn seeds germinate, they tend to grow 5 inches in the first week and then 3 inches per week for the remainder of the season. The relationship between the height (H) and the number of weeks since germination (W) is shown below.

a. Complete the missing values in the table.

Number of Weeks Since Germination (W)	Height of Corn Plant (H)
1	5
2	8
3	11
4	14
5	17
6	20

b. The expression for this height is $2 + 3W$. How tall will the corn plant be after 15 weeks of growth?

$2 + 3(15) = 47$. The plant will be 47 inches tall.

6. The Honeymoon Charter Fishing Boat Company only allows newlywed couples on their sunrise trips. There is a captain, a first mate, and a deck hand manning the boat on these trips.

a. Write an expression that shows the number of people on the boat when there are C couples booked for the trip.

$3 + 2C$

b. If the boat can hold a maximum of 20 people, how many couples can go on the sunrise fishing trip?

Eight couples (16 passengers) can fit along with the 3 crew members, totaling 19 people on the boat. A ninth couple would overload the boat.

Lesson 22: Writing and Evaluating Expressions—Exponents

Student Outcomes

- Students evaluate and write formulas involving exponents for given values in real-world problems.

Lesson Notes

Exponents are used in calculations of both area and volume. Other examples of exponential applications involve bacterial growth (powers of 2) and compound interest.

Students need a full-size sheet of paper ($8\frac{1}{2} \times 11$ inches) for the first example. Teachers should try the folding activity ahead of time to anticipate outcomes. If time permits at the end of the lesson, a larger sheet of paper can be used to experiment further.

Classwork

Fluency Exercise (10 minutes): Multiplication of Decimals

RWBE: Refer to the Rapid White Board Exchanges sections in the Module Overview for directions on how to administer an RWBE.

Example 1 (5 minutes): Folding Paper

Ask students to predict how many times they can fold a piece of paper in half. Allow a short discussion before allowing students to try it.

- Predict how many times you can fold a piece of paper in half. The folds must be as close to a half as possible. Record your prediction in Exercise 1.

Students repeatedly fold a piece of paper until it is impossible, about seven folds. Remind students they must fold the paper the same way each time.

- Fold the paper once. Record the number of layers of paper that result in the table in Exercise 2.
  - 2
- Fold again. Record the number of layers of paper that result.
  - 4

Ensure that students see that doubling the two sheets results in four sheets. At this stage, the layers can easily be counted. During subsequent stages, it is impractical to do so. Focus the count on the corner that has four loose pieces.

- Fold again. Count and record the number of layers you have now.
  - 8

The number of layers is doubling from one stage to the next; so, the pattern is modeled by multiplying by 2, not adding 2. It is critical that students find that there are eight layers here, not six.

> *Scaffolding:*
>
> Some students benefit from unfolding and counting rectangles on the paper throughout Example 1. This provides a concrete representation of the exponential relationship at the heart of this lesson.

©2015 Great Minds. eureka-math.org
G6-M4-TE-B4-1.3.1-01.2016

- Continue folding and recording the number of layers you make. Use a calculator if desired. Record your answers as both numbers in standard form and exponential form, as powers of 2.

Exercises (5 minutes)

Exercises

1. Predict how many times you can fold a piece of paper in half.

    My prediction: _____

2. Before any folding (zero folds), there is only one layer of paper. This is recorded in the first row of the table.

    Fold your paper in half. Record the number of layers of paper that result. Continue as long as possible.

Number of Folds	Number of Paper Layers That Result	Number of Paper Layers Written as a Power of 2
0	1	2^0
1	2	2^1
2	4	2^2
3	8	2^3
4	16	2^4
5	32	2^5
6	64	2^6
7	128	2^7
8	256	2^8

    a. Are you able to continue folding the paper indefinitely? Why or why not?

       No. The stack got too thick on one corner because it kept doubling each time.

    b. How could you use a calculator to find the next number in the series?

       I could multiply the number by 2 to find the number of layers after another fold.

    c. What is the relationship between the number of folds and the number of layers?

       As the number of folds increases by one, the number of layers doubles.

    d. How is this relationship represented in exponential form of the numerical expression?

       I could use 2 as a base and the number of folds as the exponent.

    e. If you fold a paper f times, write an expression to show the number of paper layers.

       There would be 2^f layers of paper.

3. If the paper were to be cut instead of folded, the height of the stack would double at each successive stage, and it would be possible to continue.

    a. Write an expression that describes how many layers of paper result from 16 cuts.

       2^{16}

    b. Evaluate this expression by writing it in standard form.

       $2^{16} = 65,536$

Lesson 22:   Writing and Evaluating Expressions—Exponents

Example 2 (10 minutes): Bacterial Infection

- ▪ Modeling of exponents in real life leads to our next example of the power of doubling. Think about the last time you had a cut or a wound that became infected. What caused the infection?
  - ▫ *Bacteria growing in the wound*
- ▪ When colonies of certain types of bacteria are allowed to grow unchecked, serious illness can result.

Example 2: Bacterial Infection

Bacteria are microscopic single-celled organisms that reproduce in a couple of different ways, one of which is called *binary fission*. In binary fission, a bacterium increases its size until it is large enough to split into two parts that are identical. These two grow until they are both large enough to split into two individual bacteria. This continues as long as growing conditions are favorable.

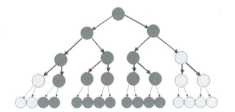

a. Record the number of bacteria that result from each generation.

Generation	Number of Bacteria	Number of Bacteria Written as a Power of 2
1	2	2^1
2	4	2^2
3	8	2^3
4	16	2^4
5	32	2^5
6	64	2^6
7	128	2^7
8	256	2^8
9	512	2^9
10	$1,024$	2^{10}
11	$2,048$	2^{11}
12	$4,096$	2^{12}
13	$8,192$	2^{13}
14	$16,384$	2^{14}

b. How many generations would it take until there were over one million bacteria present?

20 generations will produce more than one million bacteria. $2^{20} = 1,048,576$

c. Under the right growing conditions, many bacteria can reproduce every 15 minutes. Under these conditions, how long would it take for one bacterium to reproduce itself into more than one million bacteria?

It would take 20 *fifteen-minute periods, or* 5 *hours.*

d. Write an expression for how many bacteria would be present after g generations.

There will be 2^g *bacteria present after g generations.*

Example 3 (10 minutes): Volume of a Rectangular Solid

- Exponents are used when we calculate the volume of rectangular solids.

Example 3: Volume of a Rectangular Solid

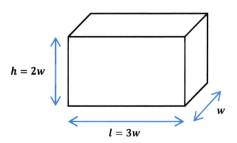

$h = 2w$

$l = 3w$

w

This box has a width, w. The height of the box, h, is twice the width. The length of the box, l, is three times the width. That is, the width, height, and length of a rectangular prism are in the ratio of $1:2:3$.

For rectangular solids like this, the volume is calculated by multiplying length times width times height.

$$V = l \cdot w \cdot h$$
$$V = 3w \cdot w \cdot 2w$$
$$V = 3 \cdot 2 \cdot w \cdot w \cdot w$$
$$V = 6w^3$$

Follow the above example to calculate the volume of these rectangular solids, given the width, w.

Width in Centimeters (cm)	Volume in Cubic Centimeters (cm^3)
1	$1 \text{ cm} \times 2 \text{ cm} \times 3 \text{ cm} = 6 \text{ cm}^3$
2	$2 \text{ cm} \times 4 \text{ cm} \times 6 \text{ cm} = 48 \text{ cm}^3$
3	$3 \text{ cm} \times 6 \text{ cm} \times 9 \text{ cm} = 162 \text{ cm}^3$
4	$4 \text{ cm} \times 8 \text{ cm} \times 12 \text{ cm} = 384 \text{ cm}^3$
w	$w \text{ cm} \times 2w \text{ cm} \times 3w \text{ cm} = 6 \, w^3 \text{ cm}^3$

MP.3

Closing (2 minutes)

- Why is 5^3 different from 5×3?
    - *5^3 means $5 \times 5 \times 5$. Five is the factor that will be multiplied by itself 3 times. That equals 125.*
    - *On the other hand, 5×3 means $5 + 5 + 5$. Five is the addend that will be added to itself 3 times. This equals 15.*

Exit Ticket (3 minutes)

Name _____   Date _____

Lesson 22: Writing and Evaluating Expressions—Exponents

Exit Ticket

1. Naomi's allowance is $2.00 per week. If she convinces her parents to double her allowance each week for two months, what will her weekly allowance be at the end of the second month (week 8)?

Week Number	Allowance
1	$2.00
2	
3	
4	
5	
6	
7	
8	
w	

2. Write the expression that describes Naomi's allowance during week w in dollars.

©2015 Great Minds. eureka-math.org
G6-M4-TE-B4-1.3.1-01.2016

Exit Ticket Sample Solutions

1. Naomi's allowance is $2.00 per week. If she convinces her parents to double her allowance each week for two months, what will her weekly allowance be at the end of the second month (week 8)?

Week Number	Allowance
1	$2.00
2	$4.00
3	$8.00
4	$16.00
5	$32.00
6	$64.00
7	$128.00
8	$256.00
w	$\$2^w$

2. Write the expression that describes Naomi's allowance during week w in dollars.

   $\$2^w$

Problem Set Sample Solutions

1. A checkerboard has 64 squares on it.

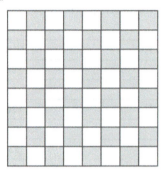

a. If one grain of rice is put on the first square, 2 grains of rice on the second square, 4 grains of rice on the third square, 8 grains of rice on the fourth square, and so on (doubling each time), complete the table to show how many grains of rice are on each square. Write your answers in exponential form on the table below.

Checkerboard Square	Grains of Rice	Checkerboard Square	Grains of Rice	Checkerboard Square	Grains of Rice	Checkerboard Square	Grains of Rice
1	2^0	17	2^{16}	33	2^{32}	49	2^{48}
2	2^1	18	2^{17}	34	2^{33}	50	2^{49}
3	2^2	19	2^{18}	35	2^{34}	51	2^{50}
4	2^3	20	2^{19}	36	2^{35}	52	2^{51}
5	2^4	21	2^{20}	37	2^{36}	53	2^{52}
6	2^5	22	2^{21}	38	2^{37}	54	2^{53}
7	2^6	23	2^{22}	39	2^{38}	55	2^{54}
8	2^7	24	2^{23}	40	2^{39}	56	2^{55}
9	2^8	25	2^{24}	41	2^{40}	57	2^{56}
10	2^9	26	2^{25}	42	2^{41}	58	2^{57}
11	2^{10}	27	2^{26}	43	2^{42}	59	2^{58}
12	2^{11}	28	2^{27}	44	2^{43}	60	2^{59}
13	2^{12}	29	2^{28}	45	2^{44}	61	2^{60}
14	2^{13}	30	2^{29}	46	2^{45}	62	2^{61}
15	2^{14}	31	2^{30}	47	2^{46}	63	2^{62}
16	2^{15}	32	2^{31}	48	2^{47}	64	2^{63}

b. How many grains of rice would be on the last square? Represent your answer in exponential form and standard form. Use the table above to help solve the problem.

There would be 2^{63} or $9,223,372,036,854,775,808$ grains of rice.

c. Would it have been easier to write your answer to part (b) in exponential form or standard form?

Answers will vary. Exponential form is more concise: 2^{63}. Standard form is longer and more complicated to calculate: $9,223,372,036,854,775,808$. (In word form: nine quintillion, two hundred twenty-three quadrillion, three hundred seventy-two trillion, thirty-six billion, eight hundred fifty-four million, seven hundred seventy-five thousand, eight hundred eight.)

2. If an amount of money is invested at an annual interest rate of 6%, it doubles every 12 years. If Alejandra invests $500, how long will it take for her investment to reach $2,000 (assuming she does not contribute any additional funds)?

It will take 24 years. After 12 years, Alejandra will have doubled her money and will have $1,000. If she waits an additional 12 years, she will have $2,000.

3. The athletics director at Peter's school has created a phone tree that is used to notify team players in the event a game has to be canceled or rescheduled. The phone tree is initiated when the director calls two captains. During the second stage of the phone tree, the captains each call two players. During the third stage of the phone tree, these players each call two other players. The phone tree continues until all players have been notified. If there are 50 players on the teams, how many stages will it take to notify all of the players?

It will take five stages. After the first stage, two players have been called, and 48 will not have been called. After the second stage, four more players will have been called, for a total of six; 44 players will remain uncalled. After the third stage, 2^3 players (eight) more will have been called, totaling 14; 36 remain uncalled. After the fourth stage, 2^4 more players (16) will have gotten a call, for a total of 30 players notified. Twenty remain uncalled at this stage. The fifth round of calls will cover all of them because 2^5 includes 32 more players.

Multiplication of Decimals

Progression of Exercises

1. $0.5 \times 0.5 =$

    0.25

2. $0.6 \times 0.6 =$

    0.36

3. $0.7 \times 0.7 =$

    0.49

4. $0.5 \times 0.6 =$

    0.3

5. $1.5 \times 1.5 =$

    2.25

6. $2.5 \times 2.5 =$

    6.25

7. $0.25 \times 0.25 =$

    0.0625

8. $0.1 \times 0.1 =$

    0.01

9. $0.1 \times 123.4 =$

    12.34

10. $0.01 \times 123.4 =$

    1.234

EUREKA
MATH

Mathematics Curriculum

GRADE 6

Topic G

Solving Equations

6.EE.B.5, 6.EE.B.6, 6.EE.B.7

Focus Standards:	6.EE.B.5	Understand solving an equation or inequality as a process of answering a question: which values from a specified set, if any, make the equation or inequality true? Use substitution to determine whether a given number in a specified set makes an equation or inequality true.
	6.EE.B.6	Use variables to represent numbers and write expressions when solving a real-world or mathematical problem; understand that a variable can represent an unknown number, or, depending on the purpose at hand, any number in a specified set.
	6.EE.B.7	Solve real-world and mathematical problems by writing and solving equations of the form $x + p = q$ and $px = q$ for cases in which p, q, and x are all nonnegative rational numbers.
Instructional Days:	7	
Lessons 23–24:	True and False Number Sentences (P, P)[1]	
Lesson 25:	Finding Solutions to Make Equations True (P)	
Lesson 26:	One-Step Equations—Addition and Subtraction (M)	
Lesson 27:	One-Step Equations—Multiplication and Division (E)	
Lesson 28:	Two-Step Problems—All Operations (M)	
Lesson 29:	Multi-Step Problems—All Operations (P)	

In Topic G, students move from identifying true and false number sentences to making true number sentences false and false number sentences true. In Lesson 23, students explain what equality and inequality symbols represent. They determine if a number sentence is true or false based on the equality or inequality symbol.

---

[1]Lesson Structure Key: **P**-Problem Set Lesson, **M**-Modeling Cycle Lesson, **E**-Exploration Lesson, **S**-Socratic Lesson

Symbol	Meaning	Example
$=$	Is equal to	$1.8 + 3 = 4.8$
$\neq$	Is not equal to	$6 \div \dfrac{1}{2} \neq 3$
$>$	Is greater than	$1 > 0.9$
$<$	Is less than	$\dfrac{1}{4} < \dfrac{1}{2}$

In Lesson 24, students move to identifying a value or a set of values that makes number sentences true. They identify values that make a true sentence false. For example, students substitute 4 for the variable in $x + 12 = 14$ to determine if the sentence is true or false. They note that when 4 is substituted for x, the sum of $x + 12$ is 16, which makes the sentence false because $16 \neq 14$. They change course in the lesson to find what they can do to make the sentence true. They ask themselves, "What number must we add to 12 to find the sum of 14?" By substituting 2 for x, the sentence becomes true because $x + 12 = 14$, $2 + 12 = 14$, and $14 = 14$. They bridge this discovery to Lesson 25 where students understand that the solution of an equation is the value or values of the variable that make the equation true.

Students begin solving equations in Lesson 26. They use bar models or tape diagrams to depict an equation and apply previously learned properties of equality for addition and subtraction to solve the equation. Students check to determine if their solutions make the equation true. Given the equation $1 + a = 6$, students represent the equation with the following model:

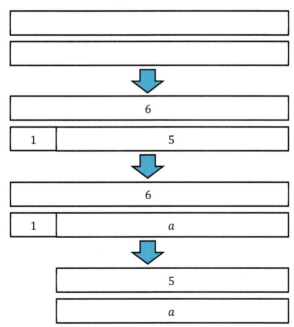

Students recognize that the solution can also be found using properties of operations. They make connections to the model and determine that $1 + a - 1 = 6 - 1$ and, ultimately, that $a = 5$. Students represent two-step and multi-step equations involving all operations with bar models or tape diagrams while continuing to apply properties of operations and the order of operations to solve equations in the remaining lessons in this topic.

EUREKA
MATH™

Lesson 23: True and False Number Sentences

Student Outcomes

- Students explain what the equality and inequality symbols including $=$, $<$, $>$, $\leq$, and $\geq$ represent. They determine if a number sentence is true or false based on the given symbol.

Lesson Notes

For students to be prepared to solve equations later in this module, it is important that they understand truth values in number sentences and in equations. In the next three lessons, students learn to differentiate between number sentences and generic equations. Later, in Lesson 26, students learn why number sentences play a fundamental role in understanding both equations with unknown numbers and solution sets of equations that contain variables. Number sentences are special among types of equations because they have truth values. A number sentence is the most concrete version of an equation. It has the very important property that it is always true or always false, and it is this property that distinguishes it from a generic equation. Examples include $2 + 3 = 5$ (true) and $2 + 2 = 5$ (false). The property guarantees the ability to check whether or not a number is a solution to an equation with a variable. Just substitute the number into the variable; then, check to see if the resulting *number sentence* is either true or false. If the number sentence is true, the number is a solution. For that reason, number sentences are the first and most important type of equation that students need to understand. Lesson 23 begins by first determining what is true and false and then moving to true and false number sentences using equality and inequality symbols.

Classwork

Opening (4 minutes)

Discuss the meaning of *true* and *false* by calling on students to tell if the following statements are true or false. Conclude with a discussion of what makes a number sentence true or false.

- Earth orbits the sun.
  - *True*
- George Washington was the first president of the United States.
  - *True*
- There are 25 hours in a day.
  - *False*
- $3 + 3 = 6$
  - *True*
- $2 + 2 = 5$
  - *False*
- Why is $2 + 2 = 5$ a false number sentence?
  - *Answers will vary but should include the idea that the expressions on both sides of the equal sign should evaluate to the same number; so, for this number sentence to be true, either the first or second addend should be three, resulting in the sum of five.*

Opening Exercise/Discussion (8 minutes)

Discuss what each symbol below stands for, and provide students with an example. Students can complete the table in their student materials.

Have one student come up to the front of the room and stand against the wall or board. Mark the student's height with a piece of masking tape. (It is important to pick a student who is about average height.) Measure the height of the tape mark using a tape measure, and record it as a mixed number in feet, along with the student's name, on the piece of masking tape. Next, start the following table on the board/screen:

Opening Exercise		
Determine what each symbol stands for, and provide an example.		
Symbol	**What the Symbol Stands For**	**Example**
$=$	*is equal to*	$4\frac{7}{8} = 4.875$

- What is another example of a number sentence that includes an equal symbol?
  - *Answers will vary. Ask more than one student.*

The student's height is the height marked by the tape on the wall. Have students stand next to the marked height. Discuss how their heights compare to the height of the tape. Are there other students in the room who have the same height?

$>$	*is greater than*	$5\frac{1}{4} > 4\frac{7}{8}$

Use the student's height measurement in the example (the example uses a student $4\frac{7}{8}$ ft. in height).

- What is another example of a number sentence that includes a greater than symbol?
  - *Answers will vary. Ask more than one student.*

Have students taller than the tape on the wall stand near the tape. Discuss how more than one student has a height that is greater than the tape, so there could be more than one number inserted into the inequality: $> 4\frac{7}{8}$.

$<$	*is less than*	$4\frac{1}{2} < 4\frac{7}{8}$

- What is another example of a number sentence that includes a less than symbol?
  - *Answers will vary. Ask more than one student.*

Have students shorter than the tape on the wall stand near the tape. Discuss how more than one student has a height that is less than the tape, so there could be more than one number inserted into the inequality: $< 4\frac{7}{8}$.

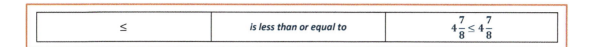

| $\leq$ | *is less than or equal to* | $4\dfrac{7}{8} \leq 4\dfrac{7}{8}$ |

- What is another example of a number sentence that includes a less than or equal to symbol?
  - *Answers will vary. Ask more than one student.*

Ask students who are the exact height as the tape and students who are shorter than the tape to stand near the tape. Discuss how this symbol is different from the previous symbol.

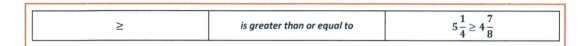

| $\geq$ | *is greater than or equal to* | $5\dfrac{1}{4} \geq 4\dfrac{7}{8}$ |

- What is another example of a number sentence that includes a greater than or equal to symbol?
  - *Answers will vary. Ask more than one student.*
- Which students would stand near the tape to demonstrate this symbol?
  - *Students who are the same height as or taller than the tape*

Example 1 (5 minutes)

Display each of the equations and inequalities one by one for students to review and determine whether they result in a true or a false number sentence.

> **Example 1**
>
> For each equation or inequality your teacher displays, write the equation or inequality, and then substitute 3 for every x. Determine if the equation or inequality results in a true number sentence or a false number sentence.

Display $5 + x = 8$.

- Substitute 3 for x and evaluate. Does this result in a true number sentence or a false number sentence?
  - *True*
- Why is the number sentence a true number sentence?
  - *Each expression on either side of the equal sign evaluates to 8. $8 = 8$*

Display $5x = 8$.

- Substitute 3 for x and evaluate. Does this result in a true number sentence or a false number sentence?
  - *False*
- Why is the number sentence a false number sentence?
  - *Five times three equals fifteen. Fifteen does not equal eight, so the number sentence $5(3) = 8$ is false.*

Display $5 + x > 8$.

- Substitute 3 for x and evaluate. Does this result in a true number sentence or a false number sentence?
  - *False*

- Why is the number sentence a false number sentence?
    - *Each expression on either side of the inequality sign evaluates to 8. However, the inequality sign states that eight is greater than eight, which is not true. We have already shown that $8 = 8$.*

Display $5x > 8$.

- Substitute 3 for x and evaluate. Does this result in a true number sentence or a false number sentence?
    - *True*
- Why is the number sentence a true number sentence?
    - *When three is substituted, the left side of the inequality evaluates to fifteen. Fifteen is greater than eight, so the number sentence is true.*

Display $5 + x \geq 8$.

- Substitute 3 for x and evaluate. Does this result in a true number sentence or a false number sentence?
    - *True*
- Why is the number sentence a true number sentence?
    - *Each expression on either side of the inequality sign evaluates to 8. Because the inequality sign states that the expression on the left side can be greater than or equal to the expression on the right side, the number sentence is true because we have already shown that $8 = 8$.*
- Can you find a number other than three that we can substitute for x that will result in a false number sentence?
    - *Answers will vary, but any number less than three will result in a false number sentence.*

Exercises (13 minutes)

Students work on the following exercises independently. Note that students are writing complete sentences to describe what happens when a variable is substituted with a number and whether it turns the equation into a true number sentence or a false number sentence.

MP.6

Exercises

Substitute the indicated value into the variable, and state (in a complete sentence) whether the resulting number sentence is true or false. If true, find a value that would result in a false number sentence. If false, find a value that would result in a true number sentence.

1. $4 + x = 12$. Substitute 8 for x.

    When 8 is substituted for x, the number sentence is true.

    Answers will vary on values to make the sentence false; any number other than 8 will make the sentence false.

2. $3g > 15$. Substitute $4\frac{1}{2}$ for g.

    When $4\frac{1}{2}$ is substituted for g, the number sentence is false.

    Answers will vary on values that make the sentence true; any number greater than 5 will make the sentence true.

Lesson 23: True and False Number Sentences

3. $\frac{f}{4} < 2$. Substitute 8 for f.

When 8 is substituted for f, the number sentence is false.

Answers will vary on values to make the sentence true; any number less than 8 will make the sentence true.

4. $14.2 \le h - 10.3$. Substitute 25.8 for h.

When 25.8 is substituted for h, the number sentence is true.

Answers will vary on values to make the sentence false; any number less than 24.5 will make the sentence false.

5. $4 = \frac{8}{h}$. Substitute 6 for h.

When 6 is substituted for h, the number sentence is false.

2 is the only value that will make the sentence true.

6. $3 > k + \frac{1}{4}$. Substitute $1\frac{1}{2}$ for k.

When $1\frac{1}{2}$ is substituted for k, the number sentence is true.

Answers will vary on values to make the sentence false; the number $2\frac{3}{4}$ or any number greater than $2\frac{3}{4}$ will make the sentence false.

7. $4.5 - d > 2.5$. Substitute 2.5 for d.

When 2.5 is substituted for d, the number sentence is false.

Answers will vary on values to make the sentence true; any number less than 2 will make the number sentence true.

8. $8 \ge 32p$. Substitute $\frac{1}{2}$ for p.

When $\frac{1}{2}$ is substituted for p, the number sentence is false.

Answers will vary on values to make the sentence true; the number $\frac{1}{4}$ or any number less than $\frac{1}{4}$ will make the sentence true.

9. $\frac{w}{2} < 32$. Substitute 16 for w.

When 16 is substituted for p, the number sentence is true.

Answers will vary on values to make the sentence false; the number 64 or any other number greater than 64 will make the sentence false.

10. $18 \le 32 - b$. Substitute 14 for b.

When 14 is substituted for b, the number sentence is true.

Answers will vary on values to make the sentence false; any number greater than 14 will make the sentence false.

©2015 Great Minds. eureka-math.org
G6-M4-TE-B4-1.3.1-01.2016

Closing (10 minutes)

Take 3 minutes to discuss the answers from the exercises. From the exercises, continue the discussion from the lesson.

- Let's take a look at Exercise 1: $4 + x = 12$. We substituted 8 for x. What did we determine?

  □ *When we substituted 8 for x, the number sentence was true.*

- And then we tried to find values to substitute for x to make the number sentence false. What number did you substitute for x to make this number sentence false?

Elicit responses from the class. Answers will vary, but collect all that make the number sentence false, and record them on the board. Elicit responses for the next set of questions:

- Did anyone substitute with zero? A thousand? A trillion? How about a fraction? A decimal?

  □ *Answers will vary.*

- If all of these responses result in a false number sentence, what can you conclude?

  □ *Only one number can be substituted to make the two expressions equal, and that number is 8.*

- Look at all of the numbers that will make this number sentence false, and then look at the one number that will make this number sentence true. Why do you think the number 8 is important compared to all the others that make the number sentence false?

Elicit various responses. The goal is for students to understand that since all numbers other than 8 result in false statements, those numbers do not contribute valuable information about the equation in the same way that the number 8 does.

- What about inequalities? Let's take another look at Exercise 2: $3g > 15$. We substituted $4\frac{1}{2}$ for g and determined that after we evaluated the inequality, it created a false number sentence because $13\frac{1}{2}$ is not greater than 15. What number did you substitute for g to make this number sentence true?

Elicit responses from the class. Answers will vary, but collect all that make the number sentence true, and record them on the board. Elicit responses for the next set of questions:

- What about 14, 16, 18, or 20? 100? 200? How about 5.1? 5.01? 5.001? 5.0000000000000001?

  □ *Answers will vary.*

- What can you conclude about the substituted numbers that will make this number sentence true?

  □ *To make this number sentence true, any number greater than five can be substituted for the variable, whether it is a whole number, fraction, or decimal.*

- Which substituted numbers made this number sentence false?

  □ *Answers will vary but must be five or any number less than five.*

- Visualize a number line in your mind. If we can only substitute numbers greater than five on the number line to make this number sentence true, what would that number line look like?

Answers will vary. The goal is for students to visualize that only part of the number line works for the inequality in order to create a true number sentence, while the other part does not work and makes the number sentence false.

> **Lesson Summary**
>
> NUMBER SENTENCE: A *number sentence* is a statement of equality (or inequality) between two numerical expressions.
>
> TRUTH VALUES OF A NUMBER SENTENCE: A number sentence is said to be *true* if both numerical expressions evaluate to the same number; it is said to be *false* otherwise. True and false are called *truth values*.
>
> Number sentences that are inequalities also have truth values. For example, $3 < 4$, $6 + 8 > 15 - 12$, and $(15 + 3)^2 < 1,000 - 32$ are all true number sentences, while the sentence $9 > 3(4)$ is false.

Exit Ticket (5 minutes)

Name _____   Date _____

Lesson 23: True and False Number Sentences

Exit Ticket

Substitute the value for the variable, and state in a complete sentence whether the resulting number sentence is true or false. If true, find a value that would result in a false number sentence. If false, find a value that would result in a true number sentence.

1. $15a \geq 75$. Substitute 5 for a.

2. $23 + b = 30$. Substitute 10 for b.

3. $20 > 86 - h$. Substitute 46 for h.

4. $32 \geq 8m$. Substitute 5 for m.

Exit Ticket Sample Solutions

Substitute the value for the variable, and state in a complete sentence whether the resulting number sentence is true or false. If true, find a value that would result in a false number sentence. If false, find a value that would result in a true number sentence.

1. $15a \geq 75$. Substitute 5 for a.

   When 5 is substituted in for a, the number sentence is true. Answers will vary, but any value for a less than 5 will result in a false number sentence.

2. $23 + b = 30$. Substitute 10 for b.

   When 10 is substituted in for b, the number sentence is false. The only value for b that will result in a true number sentence is 7.

3. $20 > 86 - h$. Substitute 46 for h.

   When 46 is substituted in for h, the number sentence will be false. Answers will vary, but any value for h greater than 66 will result in a true number sentence.

4. $32 \geq 8m$. Substitute 5 for m.

   When 5 is substituted in for m, the number sentence is false. Answers will vary, but the value of 4 and any value less than 4 for m will result in a true number sentence.

Problem Set Sample Solutions

Substitute the value for the variable, and state (in a complete sentence) whether the resulting number sentence is true or false. If true, find a value that would result in a false number sentence. If false, find a value that would result in a true number sentence.

1. $3\frac{5}{6} = 1\frac{2}{3} + h$. Substitute $2\frac{1}{6}$ for h.

   When $2\frac{1}{6}$ is substituted in for h, the number sentence is true. Answers will vary, but any value for h other than $2\frac{1}{6}$ will result in a false number sentence.

2. $39 > 156g$. Substitute $\frac{1}{4}$ for g.

   When $\frac{1}{4}$ is substituted in for g, the number sentence is false. Answers will vary, but any value for g less than $\frac{1}{4}$ will result in a true number sentence.

3. $\frac{f}{4} \leq 3$. Substitute 12 for f.

   When 12 is substituted in for f, the number sentence is true. Answers will vary, but any value for f greater than 12 will result in a false number sentence.

4. $121 - 98 \geq r$. Substitute 23 for r.

When 23 is substituted in for r, the number sentence is true. Answers will vary, but any value for r greater than 23 will result in a false number sentence.

5. $\dfrac{54}{q} = 6$. Substitute 10 for q.

When 10 is substituted in for q, the number sentence is false. The number 9 is the only value for q that will result in a true number sentence.

Create a number sentence using the given variable and symbol. The number sentence you write must be true for the given value of the variable.

6. Variable: d Symbol: $\geq$ The sentence is true when 5 is substituted for d.

7. Variable: y Symbol: $\neq$ The sentence is true when 10 is substituted for y.

8. Variable: k Symbol: $<$ The sentence is true when 8 is substituted for k.

9. Variable: a Symbol: $\leq$ The sentence is true when 9 is substituted for a.

Answers will vary for Problems 6–9.

EUREKA MATH™

Lesson 24: True and False Number Sentences

Student Outcomes

- Students identify values for the variables in equations and inequalities that result in true number sentences.
- Students identify values for the variables in equations and inequalities that result in false number sentences.

Lesson Notes

Beginning in the previous lesson and continuing here, the language used in answering questions has been carefully chosen. Responses have been purposefully elicited from students in the form of numbers, quantities, or sentences. Soon, students see that another way to report an answer to an equation or inequality *is another equation or inequality*. For example, the solution to $3x \geq 15$ can be reported as $x \geq 5$.

During this lesson, students discover that solutions and solution sets can be represented by a sentence description, leading to (or followed by) the use of equations or inequalities. This discussion provides students with knowledge to systemically solve and check one-step equations later in the module. For example, in this lesson, students transition from

"The inequality is true for any value of x that is greater than or equal to five," to
"The inequality is true when $x \geq 5$."

This transition is preparing students to understand why they rewrite complicated-looking equations and inequalities as simpler ones (such as $x = 5$ or $x \geq 5$) to describe solutions. This is an important goal in the solution process.

The $\neq$ symbol has purposefully been omitted in these lessons because it does not satisfy all of the properties listed in Tables 4 and 5 of the Common Core State Standards. However, it is a symbol that is useful and easy to understand. Its absence from the lessons does not mean that it cannot be used in class, nor should it be forgotten.

Classwork

Opening Exercise (3 minutes)

Opening Exercise

State whether each number sentence is true or false. If the number sentence is false, explain why.

a. $4 + 5 > 9$

False. $4 + 5$ is not greater than 9.

b. $3 \cdot 6 = 18$

True

c. $32 > \dfrac{64}{4}$

True

d. $78 - 15 < 68$

True

e. $22 \geq 11 + 12$

False. 22 *is not greater than or equal to* 23*.*

Students share their answers and defend their decisions for each problem.

Example 1 (10 minutes)

The teacher leads the following discussion after students complete the table below. Have students work on the first two columns alone or in groups of two, writing true or false if the number substituted for g results in a true or false number sentence.

Example 1

Write true or false if the number substituted for g results in a true or false number sentence.

Substitute g with	$4g = 32$	$g = 8$	$3g \geq 30$	$g \geq 10$	$\dfrac{g}{2} > 2$	$g > 4$	$30 \geq 38 - g$	$g \geq 8$
8	*True*	*True*	*False*	*False*	*True*	*True*	*True*	*True*
4	*False*	*False*	*False*	*False*	*False*	*False*	*False*	*False*
2	*False*	*False*	*False*	*False*	*False*	*False*	*False*	*False*
0	*False*	*False*	*False*	*False*	*False*	*False*	*False*	*False*
10	*False*	*False*	*True*	*True*	*True*	*True*	*True*	*True*

- Let's look at $4g = 32$ and $g = 8$. What do you notice happens when 8 is substituted for g in both of the equations?

    □   8 *makes both of the equations result in true number sentences.*

- What do you notice about the substitutions with 4, 2, 0, and 10?

    □   *Each of those substituted values makes the equations result in false number sentences.*

- Why do you think that happened?

    □   *Because they are both equations, we expect that only one number can be substituted for g to result in a true number sentence. In this case,* 8 *is the only number that can be substituted to make both equations true.*

- How are $4g = 32$ and $g = 8$ related?

    □   *You can get from* $4g = 32$ *to* $g = 8$ *by dividing both sides of* $4g = 32$ *by* 4*. You can get from* $g = 8$ *to* $4g = 32$ *by multiplying both sides of* $g = 8$ *by* 4*.*

- In which equation is it easier to observe the value of g that makes the number sentence true?

    □   *The second. It is certainly easier to recognize the value in the second equation.*

- Let's look at the next set of inequalities: $3g \geq 30$ and $g \geq 10$. (Let students fill out the table for these two columns.) What do you notice happens when 10 is substituted for g in both of the inequalities?

    □   10 *makes both of the inequalities result in true number sentences.*

©2015 Great Minds. eureka-math.org
G6-M4-TE-B4-1.3.1-01.2016

- Let's substitute some numbers into each inequality to test. For the second inequality, as long as the number is greater than or equal to 10, the inequality will result in a true number sentence. Let's read the inequality aloud together.
    - *Chorally: g is greater than or equal to* 10.
- Let's try 11. 11 is greater than or equal to 10. Substitute 11 for g in $3g \geq 30$. Does this result in a true number sentence?
    - *Yes*
- How are $3g \geq 30$ and $g \geq 10$ related?
    - *You can get from* $3g \geq 30$ *to* $g \geq 10$ *by dividing both sides of* $3g \geq 30$ *by* 3. *You can get from* $g \geq 10$ *to* $3g \geq 30$ *by multiplying both sides of* $g \geq 10$ *by* 3.
- In which inequality is it easier to observe the value of g that makes the number sentence true?
    - *The second, which is similar to the first example*

Continue testing the substitutions, and continue the discussion for the remaining sets of inequalities (but do not ask how the last two inequalities are related). The goal is to have students discover that for each set of equations and inequalities, the second in the set represents a *much clearer* way to represent the solutions. Point out to students that the second equation or inequality plays an important role in the next few lessons. Please note that it is not necessary that students fully understand a process for solving equations and inequalities from these examples.

Example 2 (10 minutes)

Guide students in how to use mental math to answer the questions. Students do not know how to solve one-step equations using a formal method; therefore, they need guidance in solving these problems. Please note that the second problem includes the use of subtraction to get a negative value. While operations with integers is a Grade 7 topic, this example should be accessible using a visual model.

Example 2

State when the following equations/inequalities will be true and when they will be false.

    a. $r + 15 = 25$

MP.6

- Can you think of a number that will make this equation true?
    - *Yes. Substituting* 10 *for r will make a true number sentence.*
- Is 10 the only number that results in a true number sentence? Why or why not?
    - *Yes. There is only one value that, if substituted, will result in a true number sentence. There is only one number that can be added to* 15 *to get exactly* 25.
- What will make the number sentence false?
    - *Any number that is not* 10 *will result in a false number sentence.*
- If we look back to the original questions, how can we state when the equation will be true? False?
    - *The equation is true when the value substituted for r is* 10 *and false when the value of r is any other number.*

---

    b. $6 - d > 0$

- If we wanted $6 - d$ to equal 0, what would the value of d have to be? Why?
  - *The value of d would have to be 6 because $6 - 6 = 0$.*
- Will substituting 6 for d result in a true number sentence? Why or why not?
  - *If d has a value of 6, then the resulting number sentence would not be true because the left side has to be greater than 0, not equal to 0.*
- How about substituting 5 for d? 4? 3? 2?
  - *Yes. Substituting any of these numbers for d into the inequality results in true number sentences.*
- What values can we substitute for d in order for the resulting number sentence to be true?
  - *The inequality is true for any value of d that is less than 6.*
- What values for d would make the resulting number sentence false?
  - *The inequality is false for any value of d that is greater than or equal to 6.*
- Let's take a look at a number line and see why these statements make sense.

Display a number line on the board. Label the number line as shown below.

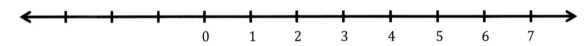

- Let's begin at 6. If I were to subtract 1 from 6, where would that place be on the number line?
  - 5
- So, if we substitute 1 for d, then $6 - 1 = 5$, and the resulting number sentence is true. How about if I subtracted 2 from 6? Would our number sentence be true for the value 2?
  - *Yes*
- What if I subtracted 6 from the 6 on the number line? Where would that be on the number line?
  - 0
- So, if we substitute 6 for d, will the resulting number sentence be true or false?
  - *False*
- Let's try one more. We have already determined that any number greater than or equal to 6 will result in a false number sentence. Let's try a number greater than 6. Let's try the number 7.
- Start with the 6 on the number line. If we were to subtract 7, in which direction on the number line would we move?
  - *To the left*
- And how many times will we move to the left?
  - 7

Model beginning at 6 on the number line, and move a finger, or draw the unit skips, while continually moving to the left on the number line 7 times.

- So, it seems we have ended up at a place to the left of 0. What number is represented by this position?
  - -1

EUREKA
MATH™

Label the number line with -1.

- Using our knowledge of ordering rational numbers, is -1 greater than or less than 0?
  - *Less than*
- So, we have shown that the inequality is true for any value of d that is less than 6 ($d < 6$) and is false when the value of d is greater than or equal to 6 ($d \geq 6$).

Continue to discuss how to answer each question below with students. As students gain more confidence, have them try to solve the problems individually; discuss the answers when students are finished.

> c. $\frac{1}{2}f = 15$
>
> *The equation is true when the value substituted for f is 30 (f = 30) and false when the value of f is any other number (f ≠ 30).*
>
> d. $\frac{y}{3} < 10$
>
> *The inequality is true for any value of y that is less than 30 (y < 30) and false when the value of y is greater than or equal to 30 (y ≥ 30).*
>
> e. $7g \geq 42$
>
> *The inequality is true for any value of g that is greater than or equal to 6 (g ≥ 6) and false when the value of g is less than (g < 6).*
>
> f. $a - 8 \leq 15$
>
> *The inequality is true for any value of a that is less than or equal to 23 (a ≤ 23) and false when the value of a is greater than 23 (a > 23).*

MP.6

Exercises (10 minutes)

Students complete the following problems in pairs.

> **Exercises**
>
> Complete the following problems in pairs. State when the following equations and inequalities will be true and when they will be false.
>
> 1. $15c > 45$
>
>   *The inequality is true for any value of c that is greater than 3 (c > 3) and false when the value of c is less than or equal to (c ≤ 3).*
>
> 2. $25 = d - 10$
>
>   *The equation is true when the value of d is 35 (d = 35) and false when the value of d is any other number (d ≠ 35).*
>
> 3. $56 \geq 2e$
>
>   *The inequality is true for any value of e that is less than or equal to 28 (e ≤ 28) and false when the value of e is greater than 8 (e > 28).*

4. $\dfrac{h}{5} \geq 12$

   The inequality is true for any value of h that is greater than or equal to 60 (h ≥ 60) and false when the value of h is less than (h < 60).

5. $45 > h + 29$

   The inequality is true for any value of h that is less than 16 (h < 16) and false when the value of h is greater than or equal to 16 (h ≥ 16).

6. $4a \leq 16$

   The inequality is true for any value of a that is less than or equal to 4 (a ≤ 4) and false when the value of a is greater than (a > 4).

7. $3x = 24$

   The equation is true when the value of x is 8 (x = 8) and false when the value of x is any other number (x ≠ 8).

Identify all equality and inequality signs that can be placed into the blank to make a true number sentence.

MP.6

8. $15 + 9$ _____ 24

   = or ≥ or ≤

9. $8 \cdot 7$ _____ 50

   > or ≥

10. $\dfrac{15}{2}$ _____ 10

    < or ≤

11. 34 _____ $17 \cdot 2$

    = or ≥ or ≤

12. 18 _____ $24.5 - 6$

    < or ≤

Lesson 24:   True and False Number Sentences

EUREKA
MATH

Closing (7 minutes)

- For the past two lessons, we have been using sentences to describe when values substituted for variables in equations and inequalities result in true number sentences or false number sentences.

- Let's take a look at an example from each of the past two lessons.

Display the following equation on the board: $5 + x = 8$.

- Substituting 3 for x in the equation results in a true number sentence: $5 + 3 = 8$. Let's evaluate to be sure.

- What is the sum of $5 + 3$?

    ▫ 8

- Does $8 = 8$?

    ▫ *Yes*

- So, when we substitute 3 for x, the equation results in a true number sentence. Let's try to substitute 3 for x in $x = 3$.

Display $x = 3$ on the board.

- If we substituted 3 for x, what would our number sentence look like?

    ▫ $3 = 3$

- Is this a true number sentence?

    ▫ *Yes*

- Previously, we described the values of x that would make the equation $5 + x = 8$ true in a sentence.

Display on the board: The equation is true when the value of x is 3.

- This is the same sentence we would write for the equation $x = 3$. Therefore, we can shorten this sentence and, instead, say: The equation is true *when $x = 3$.*

Display on the board: The equation is true when $x = 3$.

- Let's look at an inequality from today:

Display $4a \leq 16$ on the board.

- What numbers did we determine would make this inequality result in a true number sentence?

    ▫ *We determined that any number less than or equal to 4 would result in a true number sentence.*

Write this statement on the board: The inequality is true for any value of a that is less than or equal to 4.

- Is there any way we can abbreviate or shorten this statement using symbols instead of words?

Display $a \leq 4$ on the board.

- Let's read this aloud: a is less than or equal to four. We can use this inequality to rewrite the sentence.

Display on the board: The inequality is true when $a \leq 4$.

- Either sentence is a correct way to state the values that make $4a \leq 16$ true.

Exit Ticket (5 minutes)

Name _____   Date _____

Lesson 24: True and False Number Sentences

Exit Ticket

State when the following equations and inequalities will be true and when they will be false.

1. $5g > 45$

2. $14 = 5 + k$

3. $26 - w < 12$

4. $32 \le a + 8$

5. $2 \cdot h \le 16$

EUREKA
MATH

Exit Ticket Sample Solutions

> **State when the following equations and inequalities will be true and when they will be false.**
>
> 1. $5g > 45$
>
>   *The inequality is true for any value of g that is greater than 9 and false when the value of g is less than or equal to 9.*
>
>   *OR*
>
>   *The inequality is true when $g > 9$ and false when $g \leq 9$.*
>
> 2. $14 = 5 + k$
>
>   *The equation is true when the value of k is 9 and false when the value of k is any other number.*
>
>   *OR*
>
>   *The equation is true when $k = 9$ and false when $k \neq 9$.*
>
> 3. $26 - w < 12$
>
>   *The inequality is true for any value of w that is greater than 14 and false when the value of w is less than or equal to 14.*
>
>   *OR*
>
>   *The inequality is true when $w > 14$ and false when $w \leq 14$.*
>
> 4. $32 \leq a + 8$
>
>   *The inequality is true for any value of a that is greater than or equal to 24 and false when the value of a is less than 24.*
>
>   *OR*
>
>   *The inequality is true when $a \geq 24$ and false when $a < 24$.*
>
> 5. $2 \cdot h \leq 16$
>
>   *The inequality is true for any value of h that is less than or equal to 8 and false when the value of h is greater than 8.*
>
>   *OR*
>
>   *The inequality is true when $h \leq 8$ and false when $h > 8$.*

Problem Set Sample Solutions

> **State when the following equations and inequalities will be true and when they will be false.**
>
> 1. $36 = 9k$
>
>   *The equation is true when the value of k is 4 and false when the value of k is any number other than 4.*
>
>   *OR*
>
>   *The equation is true when $k = 4$ and false when $k \neq 4$.*
>
> 2. $67 > f - 15$
>
>   *The inequality is true for any value of f that is less than 82 and false when the value of f is greater than or equal to 82.*
>
>   *OR*
>
>   *The inequality is true when $f < 82$ and false when $f \geq 82$.*

©2015 Great Minds. eureka-math.org
G6-M4-TE-B4-1.3.1-01.2016

3. $\dfrac{v}{9} = 3$

   The equation is true when the value of v is 27 and false when the value of v is any number other than 27.

   OR

   The equation is true when $v = 27$ and false when $v \neq 27$.

4. $10 + b > 42$

   The inequality is true for any value of b that is greater than 32 and false when the value of b is less than or equal to 32.

   OR

   The inequality is true when $b > 32$ and false when $b \leq 32$.

5. $d - 8 \geq 35$

   The inequality is true for any value of d that is greater than or equal to 43 and false when the value of d is less than 43.

   OR

   The inequality is true when $d \geq 43$ and false when $d < 43$.

6. $32f < 64$

   The inequality is true for any value of f that is less than 2 and false when the value of f is greater than or equal to 2.

   OR

   The inequality is true when $f < 2$ and false when $f \geq 2$.

7. $10 - h \leq 7$

   The inequality is true for any value of h that is greater than or equal to 3 and false when the value of h is less than 3.

   OR

   The inequality is true when $h \geq 3$ and false when $h < 3$.

8. $42 + 8 \geq g$

   The inequality is true for any value of g that is less than or equal to 50 and false when the value of g is greater than 50.

   OR

   The inequality is true when $g \leq 50$ and false when $g > 50$.

9. $\dfrac{m}{3} = 14$

   The equation is true when the value of m is 42 and false when the value of m is any number other than 42.

   OR

   The equation is true when $m = 42$ and false when $m \neq 42$.

Lesson 25: Finding Solutions to Make Equations True

Student Outcomes

- Students learn the definition of solution in the context of placing a value into a variable to see if that value makes the equation true.

Lesson Notes

In previous lessons, students used sentences and symbols to describe the values that, when substituted for the variable in an equation, resulted in a true number sentence. In this lesson, students make the transition from their previous learning (e.g., substituting numbers into equations, writing complete sentences to describe when an equation results in a true number sentence, and using symbols to reduce the wordiness of a description) to today's lesson where they identify the value that makes an equation true as a *solution*.

As they did in previous lessons, students test for solutions by substituting numbers into equations and by checking whether the resulting number sentence is true. They have already seen how equations like $x = 3$ relate to the original equation and that it is valuable to find ways to simplify equations until they are in the form of $x = $ "a number." In the next lesson, students begin to learn the formal process of "solving an equation," that is, the process of transforming the original equation to an equation of the form $x = $ "a number," where it is easy to identify the solution.

Materials: Students complete a matching game that needs to be cut out and prepared before the class period begins. Ideally, there should be 20 sets prepared, each in a separate bag, so that students may work in pairs. Specific directions for the game are below.

Classwork

Fluency Exercise (5 minutes): Division of Fractions

Sprint: Refer to the Sprints and Sprint Delivery Script sections in the Module Overview for directions on how to administer a Sprint.

Opening Exercise (5 minutes)

> **Opening Exercise**
>
> Identify a value for the variable that would make each equation or inequality into a true number sentence. Is this the only possible answer? State when the equation or inequality is true using equality and inequality symbols.
>
> a. $3 + g = 15$
>
>   *12 is the only value of g that will make the equation true. The equation is true when $g = 12$.*
>
> b. $30 > 2d$
>
>   *Answers will vary. There is more than one value of d that will make the inequality true. The inequality is true when $d < 15$.*

---

   c. $\dfrac{15}{f} < 5$

Answers will vary. There is more than one value of f that will make the inequality true. The inequality is true when $f > 3$.

   d. $42 \le 50 - m$

Answers will vary. There is more than one value of m that will make the inequality true. The inequality is true when $m \le 8$.

Example (5 minutes)

Example

Each of the following numbers, if substituted for the variable, makes one of the equations below into a true number sentence. Match the number to that equation: $3, 6, 15, 16, 44$.

   a. $n + 26 = 32$

      6

   b. $n - 12 = 32$

      44

   c. $17n = 51$

      3

   d. $4^2 = n$

      16

   e. $\dfrac{n}{3} = 5$

      15

Discussion (2 minutes)

In most of the equations we have looked at so far, the numbers we used to substitute in for the variable have resulted in true number sentences. A number or value for the variable that results in a true number sentence is special and is called a *solution to the equation*. In the example above, 6 is a solution to $n + 26 = 32$, 44 is a solution to $n - 12 = 32$, and so on.

      Lesson 25:   Finding Solutions to Make Equations True

Exercises (15 minutes)

Students work with a partner to match the equation with its solution. Please note that below are the answers for the activity. The actual game cut-out pieces are at the end of the lesson.

MP.3

$a + 14 = 36$	22	$3^3 = b$	27
$\dfrac{c}{5} = 3$	15	$d - 10 = 32$	42
$24 = e + 11$	13	$32 = 4 \cdot f$	8
$9 = \dfrac{45}{g}$	5	$43 = h - 17$	60
$1.5 + 0.5 = j$	2	$9 \cdot \dfrac{1}{3} = k$	3
$m = \dfrac{56}{8}$	7	$n = 35.5 - 9.5$	26
$p + 13\dfrac{3}{4} = 32\dfrac{3}{4}$	19	$4 = \dfrac{1}{4} q$	16
$\dfrac{63}{r} = 7$	9	$99 - u = 45$	54

Closing (8 minutes)

- Let's look at the equation $8n = 72$. We know that 9 is a value that we can substitute for n that results in a true number sentence. In previous lessons, we described this solution as "The equation is true when the value of n is 9" and noted that the equation is false when any number other than 9 is substituted for n, or when $n \neq 9$. Therefore, there is only one solution to $8n = 72$, and it is 9.

- We also saw that both statements (i.e., the numbers that make the equation true and the numbers that make it false) can be summarized with one sentence, "The equation is true when $n = 9$," because the values that make $n = 9$ true or false are the same as the values that make $8n = 72$ true or false. Thus, we can represent the solution as "The solution is 9," or $n = 9$.

- The next lesson shows the process for transforming an equation like $8n = 72$ until it is in the form $x = 9$. You have been doing this process for many years in tape diagrams and unknown angle problems, but now we describe explicitly the steps you were following.

Note that the domain of the variable is just the set of numbers from which we are looking for solutions. For example, sometimes we only want to consider integers as solutions. In those cases, the domain of the variable would be the set of integer numbers.

> **Lesson Summary**
>
> **VARIABLE:** A *variable* is a symbol (such as a letter) that is a placeholder for a number.
>
> A variable is a placeholder for "a number" that does not "vary."
>
> **EXPRESSION:** An *expression* is a numerical expression, or it is the result of replacing some (or all) of the numbers in a numerical expression with variables.
>
> **EQUATION:** An *equation* is a statement of equality between two expressions.
>
> If A and B are two expressions in the variable x, then $A = B$ is an equation in the variable x.

Teacher notes:

A common description of a variable in the U.S. is "a quantity that varies." Ask yourselves, how can a quantity vary? A less vague description of a variable is "a placeholder for a number"; this is better because it denotes a single, non-varying number.

The upside of the description of variable (and this is a point that must be made explicit to students) is that it is the user of the variable who controls what number to insert into the placeholder. Hence, it is the student who has the power to change or vary the number as he so desires. The power to vary rests with the student, not with the variable itself!

Exit Ticket (5 minutes)

©2015 Great Minds. eureka-math.org
G6-M4-TE-B4-1.3.1-01.2016

Name _____   Date _____

Lesson 25: Finding Solutions to Make Equations True

Exit Ticket

Find the solution to each equation.

1. $7f = 49$

2. $1 = \frac{r}{12}$

3. $1.5 = d + 0.8$

4. $9^2 = h$

5. $q = 45 - 19$

6. $40 = \frac{1}{2}p$

Exit Ticket Sample Solutions

Find the solution to each equation.

1. $7f = 49$

   $f = 7$

2. $1 = \dfrac{r}{12}$

   $r = 12$

3. $1.5 = d + 0.8$

   $d = 0.7$

4. $9^2 = h$

   $h = 81$

5. $q = 45 - 19$

   $q = 26$

6. $40 = \dfrac{1}{2}p$

   $p = 80$

Problem Set Sample Solutions

Find the solution to each equation.

1. $4^3 = y$

   $y = 64$

2. $8a = 24$

   $a = 3$

3. $32 = g - 4$

   $g = 36$

4. $56 = j + 29$

   $j = 27$

5. $\dfrac{48}{r} = 12$

   $r = 4$

EUREKA MATH

6. $k = 15 - 9$

    $k = 6$

7. $x \cdot \dfrac{1}{5} = 60$

    $x = 300$

8. $m + 3.45 = 12.8$

    $m = 9.35$

9. $a = 1^5$

    $a = 1$

EUREKA
MATH™

$a + 14 = 36$	22	$3^3 = b$
27	$\dfrac{c}{5} = 3$	15
$d - 10 = 32$	42	$24 = e + 11$
13	$32 = 4 \cdot f$	8
$9 = \dfrac{45}{g}$	5	$43 = h - 17$

EUREKA
MATH™

60	$1.5 + 0.5 = j$	2
$9 \cdot \dfrac{1}{3} = k$	3	$m = \dfrac{56}{8}$
7	$n = 35.5 - 9.5$	26
$p + 13\dfrac{3}{4} = 32\dfrac{3}{4}$	19	$4 = \dfrac{1}{4}q$
16	$\dfrac{63}{r} = 7$	9

©2015 Great Minds. eureka-math.org
G6-M4-TE-B4-1.3.1-01.2016

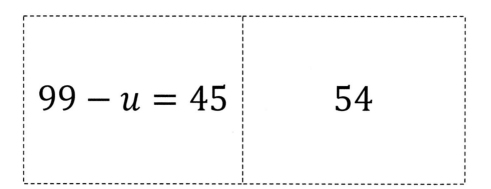

$$99 - u = 45 \quad\mid\quad 54$$

Number Correct: _____

Division of Fractions—Round 1

Directions: Evaluate each expression and simplify.

1.	9 ones ÷ 3 ones	
2.	9 ÷ 3	
3.	9 tens ÷ 3 tens	
4.	90 ÷ 30	
5.	9 hundreds ÷ 3 hundreds	
6.	900 ÷ 300	
7.	9 halves ÷ 3 halves	
8.	$\dfrac{9}{2} \div \dfrac{3}{2}$	
9.	9 fourths ÷ 3 fourths	
10.	$\dfrac{9}{4} \div \dfrac{3}{4}$	
11.	$\dfrac{9}{8} \div \dfrac{3}{8}$	
12.	$\dfrac{2}{3} \div \dfrac{1}{3}$	
13.	$\dfrac{1}{3} \div \dfrac{2}{3}$	
14.	$\dfrac{6}{7} \div \dfrac{2}{7}$	
15.	$\dfrac{5}{7} \div \dfrac{2}{7}$	
16.	$\dfrac{3}{7} \div \dfrac{4}{7}$	
17.	$\dfrac{6}{10} \div \dfrac{2}{10}$	
18.	$\dfrac{6}{10} \div \dfrac{4}{10}$	
19.	$\dfrac{6}{10} \div \dfrac{8}{10}$	
20.	$\dfrac{7}{12} \div \dfrac{2}{12}$	
21.	$\dfrac{6}{12} \div \dfrac{9}{12}$	
22.	$\dfrac{4}{12} \div \dfrac{11}{12}$	

23.	$\dfrac{6}{10} \div \dfrac{4}{10}$	
24.	$\dfrac{6}{10} \div \dfrac{2}{5} = \dfrac{6}{10} \div \dfrac{6}{10}$	
25.	$\dfrac{10}{12} \div \dfrac{5}{12}$	
26.	$\dfrac{5}{6} \div \dfrac{5}{12} = \dfrac{5}{12} \div \dfrac{5}{12}$	
27.	$\dfrac{10}{12} \div \dfrac{3}{12}$	
28.	$\dfrac{10}{12} \div \dfrac{1}{4} = \dfrac{10}{12} \div \dfrac{10}{12}$	
29.	$\dfrac{5}{6} \div \dfrac{3}{12} = \dfrac{10}{12} \div \dfrac{3}{12}$	
30.	$\dfrac{5}{10} \div \dfrac{2}{10}$	
31.	$\dfrac{5}{10} \div \dfrac{1}{5} = \dfrac{5}{10} \div \dfrac{10}{10}$	
32.	$\dfrac{1}{2} \div \dfrac{2}{10} = \dfrac{10}{10} \div \dfrac{2}{10}$	
33.	$\dfrac{1}{2} \div \dfrac{2}{4}$	
34.	$\dfrac{3}{4} \div \dfrac{2}{8}$	
35.	$\dfrac{1}{2} \div \dfrac{3}{8}$	
36.	$\dfrac{1}{2} \div \dfrac{1}{5} = \dfrac{10}{10} \div \dfrac{10}{10}$	
37.	$\dfrac{2}{4} \div \dfrac{1}{3}$	
38.	$\dfrac{1}{4} \div \dfrac{4}{6}$	
39.	$\dfrac{3}{4} \div \dfrac{2}{6}$	
40.	$\dfrac{5}{6} \div \dfrac{1}{4}$	
41.	$\dfrac{2}{9} \div \dfrac{5}{6}$	
42.	$\dfrac{5}{9} \div \dfrac{1}{6}$	
43.	$\dfrac{1}{2} \div \dfrac{1}{7}$	
44.	$\dfrac{5}{7} \div \dfrac{1}{2}$	

Division of Fractions—Round 1 [KEY]

Directions: Evaluate each expression and simplify.

1.	9 ones ÷ 3 ones	$\frac{9}{3} = 3$	23.	$\frac{6}{10} \div \frac{4}{10}$	$\frac{6}{4} = 1\frac{1}{2}$
2.	$9 \div 3$	$\frac{9}{3} = 3$	24.	$\frac{6}{10} \div \frac{2}{5} = \frac{6}{10} \div \frac{6}{10}$	$\frac{6}{4} = 1\frac{1}{2}$
3.	9 tens ÷ 3 tens	$\frac{9}{3} = 3$	25.	$\frac{10}{12} \div \frac{5}{12}$	$\frac{10}{5} = 2$
4.	$90 \div 30$	$\frac{9}{3} = 3$	26.	$\frac{5}{6} \div \frac{5}{12} = \frac{5}{12} \div \frac{5}{12}$	$\frac{10}{5} = 2$
5.	9 hundreds ÷ 3 hundreds	$\frac{9}{3} = 3$	27.	$\frac{10}{12} \div \frac{3}{12}$	$\frac{10}{3} = 3\frac{1}{3}$
6.	$900 \div 300$	$\frac{9}{3} = 3$	28.	$\frac{10}{12} \div \frac{1}{4} = \frac{10}{12} \div \frac{10}{12}$	$\frac{10}{3} = 3\frac{1}{3}$
7.	9 halves ÷ 3 halves	$\frac{9}{3} = 3$	29.	$\frac{5}{6} \div \frac{3}{12} = \frac{3}{12} \div \frac{3}{12}$	$\frac{10}{3} = 3\frac{1}{3}$
8.	$\frac{9}{2} \div \frac{3}{2}$	$\frac{9}{3} = 3$	30.	$\frac{5}{10} \div \frac{2}{10}$	$\frac{5}{2} = 2\frac{1}{2}$
9.	9 fourths ÷ 3 fourths	$\frac{9}{3} = 3$	31.	$\frac{5}{10} \div \frac{1}{5} = \frac{5}{10} \div \frac{5}{10}$	$\frac{5}{2} = 2\frac{1}{2}$
10.	$\frac{9}{4} \div \frac{3}{4}$	$\frac{9}{3} = 3$	32.	$\frac{1}{2} \div \frac{2}{10} = \frac{2}{10} \div \frac{2}{10}$	$\frac{5}{2} = 2\frac{1}{2}$
11.	$\frac{9}{8} \div \frac{3}{8}$	$\frac{9}{3} = 3$	33.	$\frac{1}{2} \div \frac{2}{4}$	$\frac{2}{2} = 1$
12.	$\frac{2}{3} \div \frac{1}{3}$	$\frac{2}{1} = 2$	34.	$\frac{3}{4} \div \frac{2}{8}$	3
13.	$\frac{1}{3} \div \frac{2}{3}$	$\frac{1}{2}$	35.	$\frac{1}{2} \div \frac{3}{8}$	$\frac{4}{3} = 1\frac{1}{3}$
14.	$\frac{6}{7} \div \frac{2}{7}$	$\frac{6}{2} = 3$	36.	$\frac{1}{2} \div \frac{1}{5} = \frac{}{10} \div \frac{}{10}$	$\frac{5}{2} = 2\frac{1}{2}$
15.	$\frac{5}{7} \div \frac{2}{7}$	$\frac{5}{2} = 2\frac{1}{2}$	37.	$\frac{2}{4} \div \frac{1}{3}$	$\frac{6}{4} = 1\frac{1}{2}$
16.	$\frac{3}{7} \div \frac{4}{7}$	$\frac{3}{4}$	38.	$\frac{1}{4} \div \frac{4}{6}$	$\frac{3}{8}$
17.	$\frac{6}{10} \div \frac{2}{10}$	$\frac{6}{2} = 3$	39.	$\frac{3}{4} \div \frac{2}{6}$	$\frac{9}{4} = 2\frac{1}{4}$
18.	$\frac{6}{10} \div \frac{4}{10}$	$\frac{6}{4} = 1\frac{1}{2}$	40.	$\frac{5}{6} \div \frac{1}{4}$	$\frac{10}{3} = 3\frac{1}{3}$
19.	$\frac{6}{10} \div \frac{8}{10}$	$\frac{6}{8} = \frac{3}{4}$	41.	$\frac{2}{9} \div \frac{5}{6}$	$\frac{4}{15}$
20.	$\frac{7}{12} \div \frac{2}{12}$	$\frac{7}{2} = 3\frac{1}{2}$	42.	$\frac{5}{9} \div \frac{1}{6}$	$\frac{15}{3} = 5$
21.	$\frac{6}{12} \div \frac{9}{12}$	$\frac{6}{9} = \frac{2}{3}$	43.	$\frac{1}{2} \div \frac{1}{7}$	$\frac{7}{2} = 3\frac{1}{2}$
22.	$\frac{4}{12} \div \frac{11}{12}$	$\frac{4}{11}$	44.	$\frac{5}{7} \div \frac{1}{2}$	$\frac{10}{7} = 1\frac{3}{7}$

Lesson 25:   Finding Solutions to Make Equations True

EUREKA MATH™

Number Correct: _____

Improvement: _____

Division of Fractions—Round 2

Directions: Evaluate each expression and simplify.

1.	12 ones ÷ 2 ones	
2.	12 ÷ 2	
3.	12 tens ÷ 2 tens	
4.	120 ÷ 20	
5.	12 hundreds ÷ 2 hundreds	
6.	1,200 ÷ 200	
7.	12 halves ÷ 2 halves	
8.	$\dfrac{12}{2} \div \dfrac{2}{2}$	
9.	12 fourths ÷ 3 fourths	
10.	$\dfrac{12}{4} \div \dfrac{3}{4}$	
11.	$\dfrac{12}{8} \div \dfrac{3}{8}$	
12.	$\dfrac{2}{4} \div \dfrac{1}{4}$	
13.	$\dfrac{1}{4} \div \dfrac{2}{4}$	
14.	$\dfrac{4}{5} \div \dfrac{2}{5}$	
15.	$\dfrac{2}{5} \div \dfrac{4}{5}$	
16.	$\dfrac{3}{5} \div \dfrac{4}{5}$	
17.	$\dfrac{6}{8} \div \dfrac{2}{8}$	
18.	$\dfrac{6}{8} \div \dfrac{4}{8}$	
19.	$\dfrac{6}{8} \div \dfrac{5}{8}$	
20.	$\dfrac{6}{10} \div \dfrac{2}{10}$	
21.	$\dfrac{7}{10} \div \dfrac{8}{10}$	
22.	$\dfrac{4}{10} \div \dfrac{7}{10}$	

23.	$\dfrac{6}{12} \div \dfrac{4}{12}$	
24.	$\dfrac{6}{12} \div \dfrac{2}{6} = \dfrac{6}{12} \div \dfrac{6}{12}$	
25.	$\dfrac{8}{14} \div \dfrac{7}{14}$	
26.	$\dfrac{8}{14} \div \dfrac{1}{2} = \dfrac{8}{14} \div \dfrac{8}{14}$	
27.	$\dfrac{11}{14} \div \dfrac{2}{14}$	
28.	$\dfrac{11}{14} \div \dfrac{1}{7} = \dfrac{11}{14} \div \dfrac{11}{14}$	
29.	$\dfrac{1}{7} \div \dfrac{6}{14} = \dfrac{6}{14} \div \dfrac{6}{14}$	
30.	$\dfrac{7}{18} \div \dfrac{3}{18}$	
31.	$\dfrac{7}{18} \div \dfrac{1}{6} = \dfrac{7}{18} \div \dfrac{7}{18}$	
32.	$\dfrac{1}{3} \div \dfrac{12}{18} = \dfrac{12}{18} \div \dfrac{12}{18}$	
33.	$\dfrac{1}{6} \div \dfrac{4}{18}$	
34.	$\dfrac{4}{12} \div \dfrac{8}{6}$	
35.	$\dfrac{1}{3} \div \dfrac{3}{15}$	
36.	$\dfrac{2}{6} \div \dfrac{1}{9} = \dfrac{}{18} \div \dfrac{}{18}$	
37.	$\dfrac{1}{6} \div \dfrac{4}{9}$	
38.	$\dfrac{2}{3} \div \dfrac{3}{4}$	
39.	$\dfrac{1}{3} \div \dfrac{3}{5}$	
40.	$\dfrac{1}{7} \div \dfrac{1}{2}$	
41.	$\dfrac{5}{6} \div \dfrac{2}{9}$	
42.	$\dfrac{5}{9} \div \dfrac{2}{6}$	
43.	$\dfrac{5}{6} \div \dfrac{4}{9}$	
44.	$\dfrac{1}{2} \div \dfrac{4}{5}$	

Division of Fractions—Round 2 [KEY]

Directions: Evaluate each expression and simplify.

1.	12 ones ÷ 2 ones	$\frac{12}{2}=6$
2.	12 ÷ 2	$\frac{12}{2}=6$
3.	12 tens ÷ 2 tens	$\frac{12}{2}=6$
4.	120 ÷ 20	$\frac{12}{2}=6$
5.	12 hundreds ÷ 2 hundreds	$\frac{12}{2}=6$
6.	1,200 ÷ 200	$\frac{12}{2}=6$
7.	12 halves ÷ 2 halves	$\frac{12}{2}=6$
8.	$\frac{12}{2}\div\frac{2}{2}$	$\frac{12}{2}=6$
9.	12 fourths ÷ 3 fourths	$\frac{12}{3}=4$
10.	$\frac{12}{4}\div\frac{3}{4}$	$\frac{12}{3}=4$
11.	$\frac{12}{8}\div\frac{3}{8}$	$\frac{12}{3}=4$
12.	$\frac{2}{4}\div\frac{1}{4}$	$\frac{2}{1}=2$
13.	$\frac{1}{4}\div\frac{2}{4}$	$\frac{1}{2}$
14.	$\frac{4}{5}\div\frac{2}{5}$	$\frac{4}{2}=2$
15.	$\frac{2}{5}\div\frac{4}{5}$	$\frac{2}{4}=\frac{1}{2}$
16.	$\frac{3}{5}\div\frac{4}{5}$	$\frac{3}{4}$
17.	$\frac{6}{8}\div\frac{2}{8}$	$\frac{6}{2}=3$
18.	$\frac{6}{8}\div\frac{4}{8}$	$\frac{6}{4}=1\frac{1}{2}$
19.	$\frac{6}{8}\div\frac{5}{8}$	$\frac{6}{5}=1\frac{1}{5}$
20.	$\frac{6}{10}\div\frac{2}{10}$	$\frac{6}{2}=3$
21.	$\frac{7}{10}\div\frac{8}{10}$	$\frac{7}{8}$
22.	$\frac{4}{10}\div\frac{7}{10}$	$\frac{4}{7}$

23.	$\frac{6}{12}\div\frac{4}{12}$	$\frac{6}{4}=1\frac{1}{2}$
24.	$\frac{6}{12}\div\frac{2}{6}=\frac{6}{12}\div\frac{6}{12}$	$\frac{6}{4}=1\frac{1}{2}$
25.	$\frac{8}{14}\div\frac{7}{14}$	$\frac{8}{7}=1\frac{1}{7}$
26.	$\frac{8}{14}\div\frac{1}{2}=\frac{8}{14}\div\frac{7}{14}$	$\frac{8}{7}=1\frac{1}{7}$
27.	$\frac{11}{14}\div\frac{2}{14}$	$\frac{11}{2}=5\frac{1}{2}$
28.	$\frac{11}{14}\div\frac{1}{7}=\frac{11}{14}\div\frac{11}{14}$	$\frac{11}{2}=5\frac{1}{2}$
29.	$\frac{1}{7}\div\frac{6}{14}=\frac{6}{14}\div\frac{6}{14}$	$\frac{2}{6}=\frac{1}{3}$
30.	$\frac{7}{18}\div\frac{3}{18}$	$\frac{7}{3}=2\frac{1}{3}$
31.	$\frac{7}{18}\div\frac{1}{6}=\frac{7}{18}\div\frac{18}{18}$	$\frac{7}{3}=2\frac{1}{3}$
32.	$\frac{1}{3}\div\frac{12}{18}=\frac{12}{18}\div\frac{18}{18}$	$\frac{6}{12}=\frac{1}{2}$
33.	$\frac{1}{6}\div\frac{4}{18}$	$\frac{3}{4}$
34.	$\frac{4}{12}\div\frac{8}{6}$	$\frac{4}{16}=\frac{1}{4}$
35.	$\frac{1}{3}\div\frac{3}{15}$	$\frac{5}{3}=1\frac{2}{3}$
36.	$\frac{2}{6}\div\frac{1}{9}=\frac{}{18}\div\frac{}{18}$	$\frac{6}{2}=3$
37.	$\frac{1}{6}\div\frac{4}{9}$	$\frac{3}{8}$
38.	$\frac{2}{3}\div\frac{3}{4}$	$\frac{8}{9}$
39.	$\frac{1}{3}\div\frac{3}{5}$	$\frac{5}{9}$
40.	$\frac{1}{7}\div\frac{1}{2}$	$\frac{2}{7}$
41.	$\frac{5}{6}\div\frac{2}{9}$	$\frac{15}{4}=3\frac{3}{4}$
42.	$\frac{5}{9}\div\frac{2}{6}$	$\frac{10}{6}=1\frac{2}{3}$
43.	$\frac{5}{6}\div\frac{4}{9}$	$\frac{15}{8}=1\frac{7}{8}$
44.	$\frac{1}{2}\div\frac{4}{5}$	$\frac{5}{8}$

Lesson 25:   Finding Solutions to Make Equations True

Lesson 26: One-Step Equations—Addition and Subtraction

Student Outcomes

- Students solve one-step equations by relating an equation to a diagram.
- Students check to determine if their solutions make the equations true.

Lesson Notes

This lesson serves as a means for students to solve one-step equations through the use of tape diagrams. Through the construction of tape diagrams, students create algebraic equations and solve for one variable. In this lesson, students continue their study of the properties of operations and identity and develop intuition of the properties of equality. This lesson continues the informal study of the properties of equality students have practiced since Grade 1 and also serves as a springboard to the formal study, use, and application of the properties of equality seen in Grade 7. While students intuitively use the properties of equality, understand that diagrams are driving the meaning behind the content of this lesson. This lesson purposefully omits focus on the actual properties of equality, which is reserved for Grade 7. Students relate an equation directly to diagrams and verbalize what they do with diagrams to construct and solve algebraic equations.

Classwork

Opening (3 minutes)

In order for students to learn how to solve multi-step equations (in future grades), they must first learn how to solve basic equations. Although a majority of students have the ability to find the solutions to the equations using mental math, it is crucial that they understand the importance of knowing and understanding the process for solving equations so they can apply it to more complex equations in the future.

Mathematical Modeling Exercise (8 minutes)

Model the example to show students how to use tape diagrams to calculate solutions to one-step equations.

Calculate the solution: $a + 2 = 8$

- Draw two tape diagrams that are the same length.

- Label the first tape diagram 8.

8

- ▪ Represent 2 on the second tape diagram. What must the remaining section of the tape diagram represent? How do you know?

	2

  - *The remaining part of the tape diagram represents 6 because the entire tape diagram is 8, and we know one section is 2. Therefore, we can compute the difference, 8 − 2, to determine the remaining part.*

- ▪ Label your tape diagram.

6	2

MP.3
&
MP.4

- ▪ Draw another set of tape diagrams to represent the given equation: $a + 2 = 8$.

8

a	2

- ▪ Because both of the following tape diagrams represent the same value, what would the value of a be? Explain.

a	2

6	2

  - *Since both of the tape diagrams represent the same value, both parts that have a and 6 must represent the same value. Therefore, a must have a value of 6.*

- ▪ Using this knowledge, try to show or explain how to solve equations without tape diagrams. What actually happened when constructing the tape diagrams?

Guide and promote this discussion with students:

  - *The first set of tape diagrams shows that the quantity of $6 + 2$ is equal to 8. To write this algebraically, we can use the equal sign. $6 + 2 = 8$*

  - *The second set of tape diagrams shows two things: first, that $a + 2$ is equal to 8, and also that $a + 2 = 8$ is equal to $6 + 2 = 8$.*

  - *We found that the only number that a can represent in the equation is 6. Therefore, when $a + 2 = 8$, the only solution for a is 6.*

- ▪ In previous lessons, we discussed identity properties. How can we explain why $a + 2 - 2 = a$ using the identity properties?

  - *We know that when we add a number and then subtract the same number, the result is the original number. Previously, we demonstrated this identity with $a + b - b = a$.*

- ▪ How can we check our answer?

  - *Substitute 6 in for a to determine if the number sentence is true. $6 + 2 = 8$ is a true number sentence because $6 + 2 - 2 = 8 - 2$, resulting in $6 = 6$. So, our answer is correct.*

Exercise 1 (8 minutes)

Students work with partners to complete the following problems. They show how to solve each equation using tape diagrams and algebraically. Then, students use substitution to check their answers after each problem.

Exercise 1

Solve each equation. Use both tape diagrams and algebraic methods for each problem. Use substitution to check your answers.

    a. $b + 9 = 15$

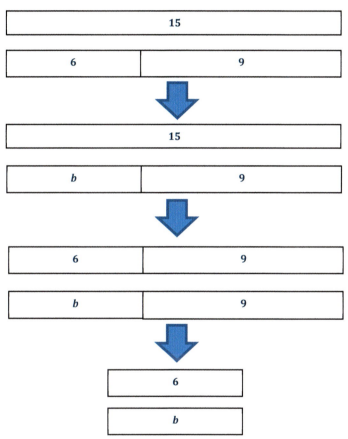

Algebraically:

$$b + 9 = 15$$
$$b + 9 - 9 = 15 - 9$$
$$b = 6$$

Check: $6 + 9 - 9 = 15 - 9$; $6 = 6$. This is a true number sentence, so 6 is the correct solution.

---

b. $12 = 8 + c$

12

8	4

12

8	c

8	4

8	c

4

c

Algebraically:

$$12 = 8 + c$$
$$12 - 8 = 8 + c - 8$$
$$4 = c$$

Check: $12 - 8 = 8 + 4 - 8$; $4 = 4$. *This is a true number sentence, so 4 is the correct solution.*

EUREKA
MATH™

Exercise 2 (8 minutes)

Students use the knowledge gained in the first part of the lesson to determine how to solve an equation with subtraction.

Exercise 2

Given the equation $d - 5 = 7$:

    a.   **Demonstrate how to solve the equation using tape diagrams.**

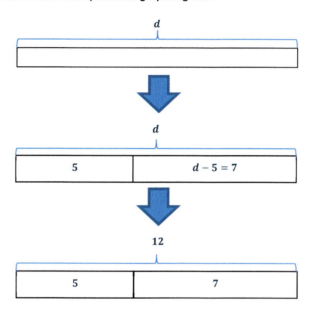

    b.   **Demonstrate how to solve the equation algebraically.**

$$d - 5 = 7$$
$$d - 5 + 5 = 7 + 5$$
$$d = 12$$

    c.   **Check your answer.**

$12 - 5 + 5 = 7 + 5; 12 = 12$. *This is a true number sentence, so our solution is correct.*

Provide students time to work and then provide some examples that show how to solve the equations using both methods. At this time, remind students of the identity with subtraction to explain why $d - 5 + 5 = d$.

Exercise 3 (8 minutes)

Students solve each problem using the method of their choice, but they must show their work. Have students check their answers.

Exercise 3

Solve each problem, and show your work. You may choose which method (tape diagrams or algebraically) you prefer. Check your answers after solving each problem.

    a. $e + 12 = 20$

20

8	12

20

e	12

8	12

e	12

8

e

Algebraically:

$$e + 12 = 20$$
$$e + 12 - 12 = 20 - 12$$
$$e = 8$$

*Check: $8 + 12 - 12 = 20 - 12; 8 = 8.$ **This is a true number sentence, so our answer is correct.***

EUREKA
MATH™

b. $f - 10 = 15$

Algebraically:

$$f - 10 = 15$$
$$f - 10 + 10 = 15 + 10$$
$$f = 25$$

Check: $25 - 10 + 10 = 15 + 10$; $25 = 25$. *This is a true number sentence, so our solution is correct.*

c. $g - 8 = 9$

Algebraically:

$$g - 8 = 9$$
$$g - 8 + 8 = 9 + 8$$
$$g = 17$$

Check: $17 - 8 + 8 = 9 + 8$; $17 = 17$. *This number sentence is true, so our solution is correct.*

Closing (5 minutes)

▪ John checked his answer and found that it was incorrect. John's work is below. What did he do incorrectly?

$$h + 10 = 25$$
$$h + 10 + 10 = 25 + 10$$
$$h = 35$$

▫ *John should have subtracted* 10 *on each side of the equation instead of adding because* $h + 10 + 10 \neq h.$

▪ Use a tape diagram to show why John's method does not lead to the correct answer.

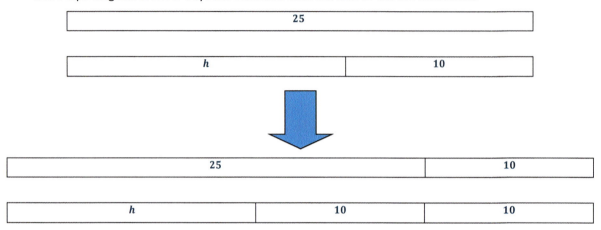

▫ *When John added* 10 *to both sides of the equation, the equation would change to* $h + 20 = 35.$ *Therefore, the value of h cannot equal* 35.

▪ Why do you do the inverse operation to calculate the solution of the equation? Include a tape diagram as part of your explanation.

▫ *When you do the inverse operation, the result is zero. Using the identity property, we know any number added to zero is the original number.*

h	10

▫ *This tape diagram demonstrates* $h + 10$; *however, we want to know the value of just h. Therefore, we would subtract* 10 *from this tape diagram.*

h

▫ *Therefore,* $h + 10 - 10 = h.$

Exit Ticket (5 minutes)

Lesson 26: One-Step Equations—Addition and Subtraction

EUREKA MATH

Name _____   Date _____

Lesson 26: One-Step Equations—Addition and Subtraction

Exit Ticket

1. If you know the answer, state it. Then, use a tape diagram to demonstrate why this is the correct answer. If you do not know the answer, find the solution using a tape diagram.

$$j + 12 = 25$$

2. Find the solution to the equation algebraically. Check your answer.

$$k - 16 = 4$$

Exit Ticket Sample Solutions

1. If you know the answer, state it. Then, use a tape diagram to demonstrate why this is the correct answer. If you do not know the answer, find the solution using a tape diagram.

$$j + 12 = 25$$

25

13	12

25

j	12

13	12

j	12

13

j

j is equal to 13; $j = 13$.

Check: $13 + 12 = 25$; $25 = 25$. This is a true number sentence, so the solution is correct.

2. Find the solution to the equation algebraically. Check your answer.

$$k - 16 = 4$$
$$k - 16 + 16 = 4 + 16$$
$$k = 20$$

Check: $20 - 16 = 4$; $4 = 4$. This is a true number sentence, so the solution is correct.

EUREKA
MATH™

Problem Set Sample Solutions

1. Find the solution to the equation below using tape diagrams. Check your answer.

$$m - 7 = 17$$

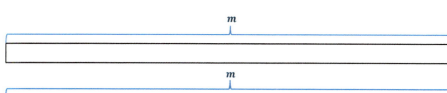

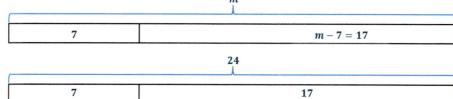

m is equal to 24; *m* = 24.

Check: 24 − 7 = 17; 17 = 17. *This number sentence is true, so the solution is correct.*

2. Find the solution of the equation below algebraically. Check your answer.

$$n + 14 = 25$$
$$n + 14 - 14 = 25 - 14$$
$$n = 11$$

Check: 11 + 14 = 25; 25 = 25. *This number sentence is true, so the solution is correct.*

3. Find the solution of the equation below using tape diagrams. Check your answer.

$$p + 8 = 18 \qquad p = 10$$

Check: 10 + 8 = 18; 18 = 18. *This number sentence is true, so the solution is correct.*

4. Find the solution to the equation algebraically. Check your answer.

$$g - 62 = 14$$
$$g - 62 + 62 = 14 + 62$$
$$g = 76$$

Check: $76 - 62 = 14$; $14 = 14$. *This number sentence is true, so the solution is correct.*

5. Find the solution to the equation using the method of your choice. Check your answer.

$$m + 108 = 243$$

Tape Diagrams:

243

135	108

243

m	108

135	108

m	108

135

m

Algebraically:

$$m + 108 = 243$$
$$m + 108 - 108 = 243 - 108$$
$$m = 135$$

Check: $135 + 108 = 243$; $243 = 243$. *This number sentence is true, so the solution is correct.*

6. Identify the mistake in the problem below. Then, correct the mistake.

$$p - 21 = 34$$
$$p - 21 - 21 = 34 - 21$$
$$p = 13$$

The mistake is subtracting rather than adding 21. *This is incorrect because* $p - 21 - 21$ *would not equal* p.

$$p - 21 = 34$$
$$p - 21 + 21 = 34 + 21$$
$$p = 55$$

Lesson 26: One-Step Equations—Addition and Subtraction

7. Identify the mistake in the problem below. Then, correct the mistake.

$$q + 18 = 22$$
$$q + 18 - 18 = 22 + 18$$
$$q = 40$$

The mistake is adding 18 on the right side of the equation instead of subtracting it from both sides.

$$q + 18 = 22$$
$$q + 18 - 18 = 22 - 18$$
$$q = 4$$

8. Match the equation with the correct solution on the right.

$$r + 10 = 22$$
$$r - 15 = 5$$
$$r - 18 = 14$$
$$r + 5 = 15$$

$$r = 10$$
$$r = 20$$
$$r = 12$$
$$r = 32$$

Lesson 27: One-Step Equations—Multiplication and Division

Student Outcomes

- Students solve one-step equations by relating an equation to a diagram.
- Students check to determine if their solutions make the equations true.

Lesson Notes

This lesson teaches students to solve one-step equations using tape diagrams. Through the construction of tape diagrams, students create algebraic equations and solve for one variable. This lesson not only allows students to continue studying the properties of operations and identity but also allows students to develop intuition of the properties of equality. This lesson continues the informal study of the properties of equality students have practiced since Grade 1 and also serves as a springboard to the formal study, use, and application of the properties of equality seen in Grade 7. Understand that, while students intuitively use the properties of equality, diagrams are the focus of this lesson. This lesson purposefully omits focusing on the actual properties of equality, which are covered in Grade 7. Students relate an equation directly to diagrams and verbalize what they do with diagrams to construct and solve algebraic equations.

Poster paper is needed for this lesson. Posters need to be prepared ahead of time, one set of questions per poster.

Classwork

Example 1 (5 minutes)

> **Example 1**
>
> Solve $3z = 9$ using tape diagrams and algebraically. Then, check your answer.
>
> First, draw two tape diagrams, one to represent each side of the equation.
>
> | 9 |
> |---|
>
> | z | z | z |
> |-----|-----|-----|
>
> If 9 had to be split into three groups, how big would each group be?
>
> 3
>
> Demonstrate the value of z using tape diagrams.
>
> | 3 | 3 | 3 |
> |---|---|---|
> | z | | |

How can we demonstrate this algebraically?

We know we have to split 9 into three equal groups, so we have to divide by 3 to show this algebraically.

$3z \div 3 = 9 \div 3$

How does this get us the value of z?

The left side of the equation will equal z because we know the identity property, where $a \cdot b \div b = a$, so we can use this identity here.

The right side of the equation will be 3 because $9 \div 3 = 3$.

Therefore, the value of z is 3.

How can we check our answer?

We can substitute the value of z into the original equation to see if the number sentence is true.

$3(3) = 9; 9 = 9.$ *This number sentence is true, so our answer is correct.*

Example 2 (5 minutes)

Example 2

Solve $\frac{y}{4} = 2$ using tape diagrams and algebraically. Then, check your answer.

First, draw two tape diagrams, one to represent each side of the equation.

$y \div 4$

2

If the first tape diagram shows the size of $y \div 4$, how can we draw a tape diagram to represent y?

The tape diagram to represent y should be four sections of the size $y \div 4$.

Draw this tape diagram.

y

$y \div 4$	$y \div 4$	$y \div 4$	$y \div 4$

What value does each $y \div 4$ section represent? How do you know?

Each $y \div 4$ section represents a value of 2. We know this from our original tape diagram.

How can you use a tape diagram to show the value of y?

Draw four equal sections of 2, which will give y the value of 8.

2	2	2	2

Lesson 27:   One-Step Equations—Multiplication and Division

289

©2015 Great Minds. eureka-math.org
G6-M4-TE-B4-1.3.1-01.2016

How can we demonstrate this algebraically?

$\frac{y}{4} \cdot 4 = 2 \cdot 4$. *Because we multiplied the number of sections in the original equation by 4, we know the identity*

$\frac{a}{b} \cdot b = a$ *can be used here.*

How does this help us find the value of y?

The left side of the equation will equal y, and the right side will equal 8. Therefore, the value of y is 8.

How can we check our answer?

Substitute 8 into the equation for y, and then check to see if the number sentence is true.

$\frac{8}{4} = 2$. *This is a true number sentence, so 8 is the correct answer.*

Exploratory Challenge (15 minutes)

Each group (two or three) of students receives one set of problems. Have students solve both problems on poster paper with tape diagrams and algebraically. Students should also check their answers on the poster paper. More than one group may have each set of problems.

> *Scaffolding:*
>
> If students are struggling, model one set of problems before continuing with the Exploratory Challenge.

Set 1

On poster paper, solve each problem below algebraically and using tape diagrams. Check each answer to show that you solved the equation correctly (algebraic and tape diagram sample responses are below).

1. $2a = 16$

    Tape Diagrams:

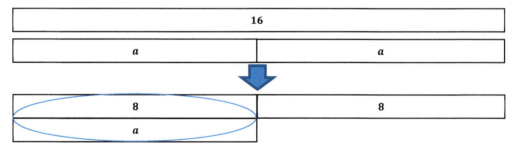

    Algebraically:

$$2a = 16$$
$$2a \div 2 = 16 \div 2$$
$$a = 8$$

Check: $2 \cdot 8 = 16; 16 = 16$. *This is a true number sentence, so 8 is the correct solution.*

MP.1

2. $\dfrac{b}{3} = 4$

    Tape Diagrams:

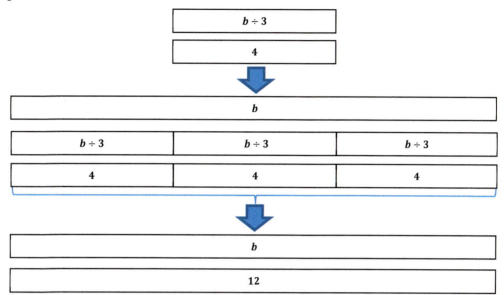

    Algebraically:

$$\frac{b}{3} = 4$$
$$\frac{b}{3} \cdot 3 = 4 \cdot 3$$
$$b = 12$$

MP.1

Check: $\dfrac{12}{3} = 4;\ 4 = 4.$ *This number sentence is true, so 12 is the correct solution.*

Set 2

On poster paper, solve each problem below algebraically and using tape diagrams. Check each answer to show that you solved the equation correctly (algebraic and tape diagram sample responses are below).

1. $4 \cdot c = 24$

    Tape Diagrams:

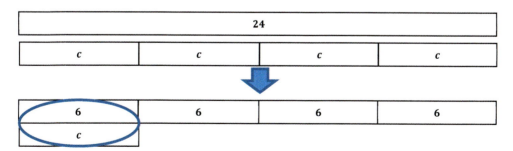

Algebraically:

$$4 \cdot c = 24$$
$$4 \cdot c \div 4 = 24 \div 4$$
$$c = 6$$

Check: $4 \cdot 6 = 24$; $24 = 24$. *This number sentence is true, so 6 is the correct solution.*

2. $\dfrac{d}{7} = 1$

Tape Diagrams:

$d \div 7$
1

d

$d \div 7$	$d \div 7$	$d \div 7$	$d \div 7$	$d \div 7$	$d \div 7$	$d \div 7$
1	1	1	1	1	1	1

d
7

Algebraically:

$$\frac{d}{7} = 1$$
$$\frac{d}{7} \cdot 7 = 1 \cdot 7$$
$$d = 7$$

Check: $\dfrac{7}{7} = 1$; $1 = 1$. *This number sentence is true, so 7 is the correct solution.*

EUREKA
MATH™

©2015 Great Minds. eureka-math.org
G6-M4-TE-B4-1.3.1-01.2016

Set 3

On poster paper, solve each problem below algebraically and using tape diagrams. Check each answer to show that you solved the equation correctly (algebraic and tape diagram sample responses are below).

1. $5e = 45$

   Tape Diagrams:

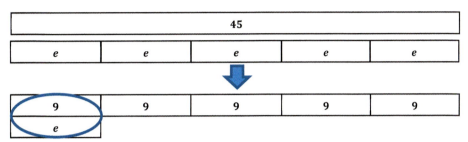

   Algebraically:

$$5e = 45$$
$$5e \div 5 = 45 \div 5$$
$$e = 9$$

   Check: $5(9) = 45; 45 = 45$. *This number sentence is true, so 9 is the correct solution.*

2. $\dfrac{f}{3} = 10$

MP.1

   Tape Diagrams:

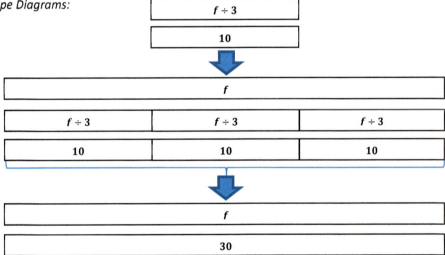

   Algebraically:

$$\frac{f}{3} = 10$$
$$\frac{f}{3} \cdot 3 = 10 \cdot 3$$
$$f = 30$$

   Check: $\dfrac{30}{3} = 10; 10 = 10$. *This number sentence is true, so 30 is the correct solution.*

Set 4

On poster paper, solve each problem below algebraically and using tape diagrams. Check each answer to show that you solved the equation correctly (algebraic and tape diagram sample responses are below).

1. $9 \cdot g = 54$

    Tape Diagrams:

54								

g	g	g	g	g	g	g	g	g

    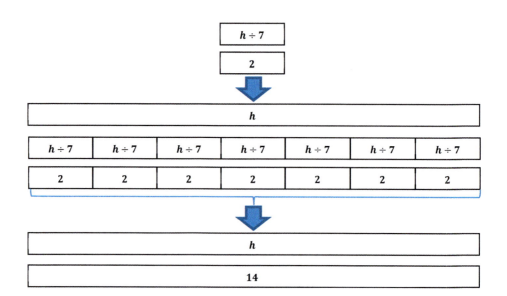

6	6	6	6	6	6	6	6	6
g

    Algebraically:

    $$9 \cdot g = 54$$
    $$9 \cdot g \div 9 = 54 \div 9$$
    $$g = 6$$

    Check: $9 \cdot 6 = 54$; $54 = 54$. *This number sentence is true, so 6 is the correct solution.*

    MP.1

2. $2 = \dfrac{h}{7}$

    Tape Diagrams:

$h \div 7$

2

h

$h \div 7$	$h \div 7$	$h \div 7$	$h \div 7$	$h \div 7$	$h \div 7$	$h \div 7$
2	2	2	2	2	2	2

h

14

EUREKA
MATH™

Algebraically:

$$2 = \frac{h}{7}$$

$$2 \cdot 7 = \frac{h}{7} \cdot 7$$

$$14 = h$$

Check: $2 = \frac{14}{7}$; $2 = 2$. *This number sentence is true, so 14 is the correct solution.*

Set 5

On poster paper, solve each problem below algebraically and using tape diagrams. Check each answer to show that you solved the equation correctly (algebraic and tape diagram sample responses are below).

1. $50 = 10j$

MP.1   *Tape Diagrams:*

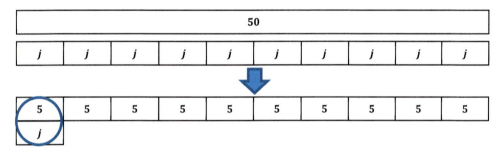

Algebraically:

$$50 = 10j$$

$$50 \div 10 = 10j \div 10$$

$$5 = j$$

Check: $50 = 10(5)$; $50 = 50$. *This number sentence is true, so 5 is the correct solution.*

2. $\dfrac{k}{8} = 3$

Tape Diagrams:

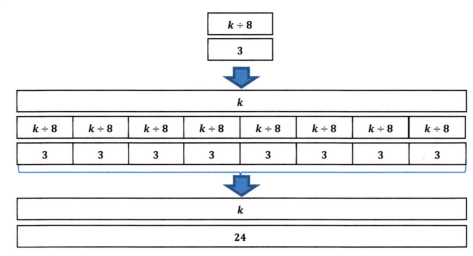

Algebraically:

$$\frac{k}{8} = 3$$

$$\frac{k}{8} \cdot 8 = 3 \cdot 8$$

$$k = 24$$

Check: $\dfrac{24}{8} = 3; \ 3 = 3.$ *This number sentence is true, so* 24 *is the correct solution.*

MP.1

MP.3 Hang completed posters around the room. Students walk around to examine other groups' posters. Students may either write on a piece of paper, write on Post-it notes, or write on the posters any questions or comments they may have. Answer students' questions after providing time for students to examine posters.

EUREKA
MATH™

Exercises (10 minutes)

Students complete the following problems individually. Remind students to check their solutions.

Exercises

1. Use tape diagrams to solve the following problem: $3m = 21$.

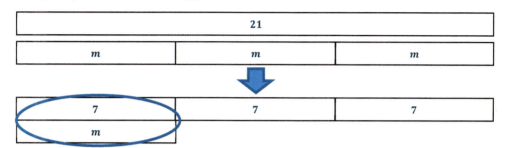

Check: $3(7) = 21$; $21 = 21$. *This number sentence is true, so 7 is the correct solution.*

2. Solve the following problem algebraically: $15 = \frac{n}{5}$.

$$15 = \frac{n}{5}$$
$$15 \cdot 5 = \frac{n}{5} \cdot 5$$
$$75 = n$$

Check: $15 = \frac{75}{5}$; $15 = 15$. *This number sentence is true, so 75 is the correct solution.*

3. Calculate the solution of the equation using the method of your choice: $4p = 36$.

Tape Diagrams:

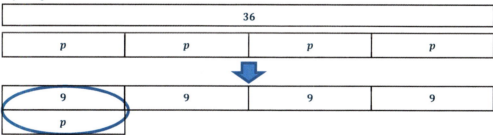

Algebraically:

$$4p = 36$$
$$4p \div 4 = 36 \div 4$$
$$p = 9$$

Check: $4(9) = 36$; $36 = 36$. *This number sentence is true, so 9 is the correct solution.*

EUREKA MATH™

4. Examine the tape diagram below, and write an equation it represents. Then, calculate the solution to the equation using the method of your choice.

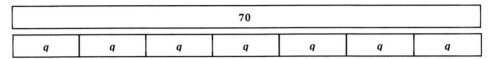

$7q = 70$ or $70 = 7q$

Tape Diagram:

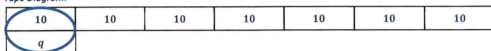

Algebraically:

$$7q = 70 \qquad\qquad 70 = 7q$$
$$7q \div 7 = 70 \div 7 \qquad 70 \div 7 = 7q \div 7$$
$$q = 10 \qquad\qquad q = 10$$

Check: $7(10) = 70$, $70 = 7(10)$; $70 = 70$. This number sentence is true, so 10 is the correct answer.

5. Write a multiplication equation that has a solution of 12. Use tape diagrams to prove that your equation has a solution of 12.

Answers will vary.

6. Write a division equation that has a solution of 12. Prove that your equation has a solution of 12 using algebraic methods.

Answers will vary.

Closing (5 minutes)

- How is solving addition and subtraction equations similar to and different from solving multiplication and division equations?
  - *Solving addition and subtraction equations is similar to solving multiplication and division equations because identities are used for all of these equations.*
  - *Solving addition and subtraction equations is different from solving multiplication and division equations because they require different identities.*
- What do you know about the pattern in the operations you used to solve the equations today?
  - *We used inverse operations to solve the equations today. Division was used to solve multiplication equations, and multiplication was used to solve division equations.*

Exit Ticket (5 minutes)

EUREKA
MATH™

Name _____   Date _____

Lesson 27: One-Step Equations—Multiplication and Division

Exit Ticket

Calculate the solution to each equation below using the indicated method. Remember to check your answers.

1. Use tape diagrams to find the solution of $\dfrac{r}{10} = 4$.

2. Find the solution of $64 = 16u$ algebraically.

3. Use the method of your choice to find the solution of $12 = 3v$.

EUREKA
MATH

Exit Ticket Sample Solutions

Calculate the solution to each equation below using the indicated method. Remember to check your answers.

1. Use tape diagrams to find the solution of $\dfrac{r}{10} = 4$.

$r \div 10$

4

r

$r \div 10$	$r \div 10$	$r \div 10$	$r \div 10$	$r \div 10$	$r \div 10$	$r \div 10$	$r \div 10$	$r \div 10$	$r \div 10$

4	4	4	4	4	4	4	4	4	4

r

40

Check: $\dfrac{40}{10} = 4$; $4 = 4$. *This number sentence is true, so 40 is the correct solution.*

2. Find the solution of $64 = 16u$ algebraically.

$$64 = 16u$$
$$64 \div 16 = 16u \div 16$$
$$4 = u$$

Check: $64 = 16(4)$; $64 = 64$. *This number sentence is true, so 4 is the correct solution.*

3. Use the method of your choice to find the solution of $12 = 3v$.

Tape Diagrams:

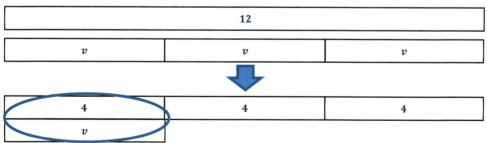

12

v	v	v

4	4	4
v		

Algebraically:

$$12 = 3v$$
$$12 \div 3 = 3v \div 3$$
$$4 = v$$

Check: $12 = 3(4)$; $12 = 12$. *This number sentence is true, so 4 is the correct solution.*

Problem Set Sample Solutions

1. Use tape diagrams to calculate the solution of $30 = 5w$. Then, check your answer.

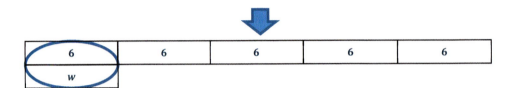

30

w	w	w	w	w

6	6	6	6	6
w				

Check: $30 = 5(6)$; $30 = 30$. *This number sentence is true, so 6 is the correct solution.*

2. Solve $12 = \frac{x}{4}$ algebraically. Then, check your answer.

$$12 = \frac{x}{4}$$
$$12 \cdot 4 = \frac{x}{4} \cdot 4$$
$$48 = x$$

Check: $12 = \frac{48}{4}$; $12 = 12$. *This number sentence is true, so 48 is the correct solution.*

3. Use tape diagrams to calculate the solution of $\frac{y}{5} = 15$. Then, check your answer.

$y \div 5$

15

y

$y \div 5$	$y \div 5$	$y \div 5$	$y \div 5$	$y \div 5$

15	15	15	15	15

y

75

Check: $\frac{75}{5} = 15$; $15 = 15$. *This number sentence is true, so 75 is the correct solution.*

EUREKA
MATH™

4. Solve $18z = 72$ algebraically. Then, check your answer.

$$18z = 72$$
$$18z \div 18 = 72 \div 18$$
$$z = 4$$

Check: $18(4) = 72; 72 = 72.$ *This number sentence is true, so 4 is the correct solution.*

5. Write a division equation that has a solution of 8. Prove that your solution is correct by using tape diagrams.

Answers will vary.

6. Write a multiplication equation that has a solution of 8. Solve the equation algebraically to prove that your solution is correct.

Answers will vary.

7. When solving equations algebraically, Meghan and Meredith each got a different solution. Who is correct? Why did the other person not get the correct answer?

Meghan	Meredith
$\dfrac{y}{2} = 4$	$\dfrac{y}{2} = 4$
$\dfrac{y}{2} \cdot 2 = 4 \cdot 2$	$\dfrac{y}{2} \div 2 = 4 \div 2$
$y = 8$	$y = 2$

Meghan is correct. Meredith divided by 2 to solve the equation, which is not correct because she would end up with $\dfrac{y}{4} = 2$. To solve a division equation, Meredith must multiply by 2 to end up with y because the identity states $y \div 2 \cdot 2 = y$.

Lesson 28: Two-Step Problems—All Operations

Student Outcomes

- Students calculate the solutions of two-step equations by using their knowledge of order of operations and the properties of equality for addition, subtraction, multiplication, and division. Students employ tape diagrams to determine their answers.

- Students check to determine if their solutions make the equations true.

Classwork

Fluency Exercise (5 minutes): Addition of Decimals

Sprint: Refer to the Sprints and Sprint Delivery Script sections in the Module Overview for directions on how to administer a Sprint.

Mathematical Modeling Exercise (6 minutes)

Model the problems while students follow along.

Mathematical Modeling Exercise

Juan has gained 20 lb. since last year. He now weighs 120 lb. Rashod is 15 lb. heavier than Diego. If Rashod and Juan weighed the same amount last year, how much does Diego weigh? Let j represent Juan's weight last year in pounds, and let d represent Diego's weight in pounds.

Draw a tape diagram to represent Juan's weight.

120	
j	20

Draw a tape diagram to represent Rashod's weight.

d	15

Draw a tape diagram to represent Diego's weight.

d

What would combining all three tape diagrams look like?

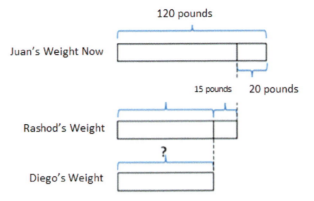

Write an equation to represent Juan's tape diagram.

$j + 20 = 120$

Write an equation to represent Rashod's tape diagram.

$d + 15 + 20 = 120$

How can we use the final tape diagram or the equations above to answer the question presented?

By combining 15 and 20 from Rashod's equation, we can use our knowledge of addition identities to determine Diego's weight.

The final tape diagram can be used to write a third equation, $d + 35 = 120$. We can use our knowledge of addition identities to determine Diego's weight.

Calculate Diego's weight.

$$d + 35 - 35 = 120 - 35$$
$$d = 85$$

We can use identities to defend our thought that $d + 35 - 35 = d$.

Does your answer make sense?

Yes. If Diego weighs 85 lb., and Rashod weighs 15 lb. more than Diego, then Rashod weighs 100 lb., which is what Juan weighed before he gained 20 lb.

Example 1 (5 minutes)

Assist students in solving the problem by providing step-by-step guidance.

Example 1

Marissa has twice as much money as Frank. Christina has $20 more than Marissa. If Christina has $100, how much money does Frank have? Let f represent the amount of money Frank has in dollars and m represent the amount of money Marissa has in dollars.

Draw a tape diagram to represent the amount of money Frank has.

f

Draw a tape diagram to represent the amount of money Marissa has.

f	f

Draw a tape diagram to represent the amount of money Christina has.

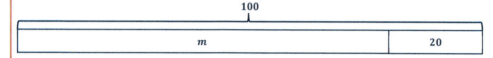

Which tape diagram provides enough information to determine the value of the variable m?

The tape diagram that represents the amount of money Christina has.

Write and solve the equation.

$$m + 20 = 100$$
$$m + 20 - 20 = 100 - 20$$
$$m = 80$$

The identities we have discussed throughout the module solidify that $m + 20 - 20 = m$.

What does the 80 represent?

80 is the amount of money, in dollars, that Marissa has.

Now that we know Marissa has $80, how can we use this information to find out how much money Frank has?

We can write an equation to represent Marissa's tape diagram since we now know the length is 80.

Write an equation.

$2f = 80$

EUREKA MATH™

Solve the equation.

$$2f \div 2 = 80 \div 2$$
$$f = 40$$

Once again, the identities we have used throughout the module can solidify that $2f \div 2 = f$.

What does the 40 represent?

The 40 represents the amount of money Frank has in dollars.

Does 40 make sense in the problem?

Yes, because if Frank has $\$40$, then Marissa has twice this, which is $\$80$. Then, Christina has $\$100$ because she has $\$20$ more than Marissa, which is what the problem stated.

Exercises (20 minutes; 5 minutes per station)

Students work in small groups to complete the following stations.

Station One: Use tape diagrams to solve the problem.

Raeana is twice as old as Madeline, and Laura is 10 years older than Raeana. If Laura is 50 years old, how old is Madeline? Let m represent Madeline's age in years, and let r represent Raeana's age in years.

Raeana's Tape Diagram:

m	m

Madeline's Tape Diagram:

m

Laura's Tape Diagram:

Equation for Laura's Tape Diagram:

$$r + 10 = 50$$
$$r + 10 - 10 = 50 - 10$$
$$r = 40$$

We now know that Raeana is 40 years old, and we can use this and Raeana's tape diagram to determine the age of Madeline.

$$2m = 40$$
$$2m \div 2 = 40 \div 2$$
$$m = 20$$

Therefore, Madeline is 20 years old.

MP.1

Station Two: Use tape diagrams to solve the problem.

Carli has 90 apps on her phone. Braylen has half the amount of apps as Theiss. If Carli has three times the amount of apps as Theiss, how many apps does Braylen have? Let b represent the number of Braylen's apps and t represent the number of Theiss's apps.

Theiss's Tape Diagram:

t

Braylen's Tape Diagram:

b	b

Carli's Tape Diagram:

90

t	t	t

Equation for Carli's Tape Diagram:

We now know that Theiss has 30 apps on his phone. We can use this information to write an equation for Braylen's tape diagram and determine how many apps are on Braylen's phone.

$$2b = 30$$
$$2b \div 2 = 30 \div 2$$
$$b = 15$$

Therefore, Braylen has 15 apps on his phone.

MP.1

Station Three: Use tape diagrams to solve the problem.

Reggie ran for 180 yards during the last football game, which is 40 more yards than his previous personal best. Monte ran 50 more yards than Adrian during the same game. If Monte ran the same amount of yards Reggie ran in one game for his previous personal best, how many yards did Adrian run? Let r represent the number of yards Reggie ran during his previous personal best and a represent the number of yards Adrian ran.

Reggie's Tape Diagram:

180

r	40

Monte's Tape Diagram:

a	50

Adrian's Tape Diagram:

a

Combining all 3 tape diagrams:

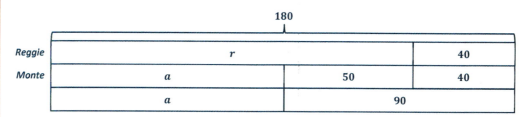

Equation for Reggie's Tape Diagram:

$$r + 40 = 180$$

Equation for Monte's Tape Diagram:

$$a + 50 + 40 = 180$$
$$a + 90 = 180$$
$$a + 90 - 90 = 180 - 90$$
$$a = 90$$

Therefore, Adrian ran 90 yards during the football game.

<u>Station Four:</u> Use tape diagrams to solve the problem.

Lance rides his bike downhill at a pace of 60 miles per hour. When Lance is riding uphill, he rides 8 miles per hour slower than on flat roads. If Lance's downhill speed is 4 times faster than his flat-road speed, how fast does he travel uphill? Let f represent Lance's pace on flat roads in miles per hour and u represent Lance's pace uphill in miles per hour.

MP.1

Tape Diagram for Uphill Pace:

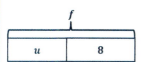

Tape Diagram for Downhill Pace:

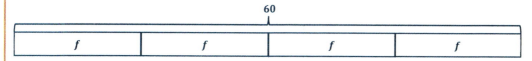

Equation for Downhill Pace:

$$4f = 60$$
$$4f \div 4 = 60 \div 4$$
$$f = 15$$

Equation for Uphill Pace:

$$u + 8 = 15$$
$$u + 8 - 8 = 15 - 8$$
$$u = 7$$

Therefore, Lance travels at a pace of 7 miles per hour uphill.

Closing (4 minutes)

Use this time to go over the solutions to the stations and answer student questions.

- How did the tape diagrams help you create the expressions and equations that you used to solve the problems?

  □ *Answers will vary.*

Exit Ticket (5 minutes)

Name _____   Date _____

Lesson 28: Two-Step Problems—All Operations

Exit Ticket

Use tape diagrams and equations to solve the problem with visual models and algebraic methods.

Alyssa is twice as old as Brittany, and Jazmyn is 15 years older than Alyssa. If Jazmyn is 35 years old, how old is Brittany? Let a represent Alyssa's age in years and b represent Brittany's age in years.

Exit Ticket Sample Solutions

Use tape diagrams and equations to solve the problem with visual models and algebraic methods.

Alyssa is twice as old as Brittany, and Jazmyn is 15 years older than Alyssa. If Jazmyn is 35 years old, how old is Brittany? Let a represent Alyssa's age in years and b represent Brittany's age in years.

Brittany's Tape Diagram:

b

Alyssa's Tape Diagram:

b	b

Jazmyn's Tape Diagram:

35

a	15

Equation for Jazmyn's Tape Diagram:

$$a + 15 = 35$$
$$a + 15 - 15 = 35 - 15$$
$$a = 20$$

Now that we know Alyssa is 20 years old, we can use this information and Alyssa's tape diagram to determine Brittany's age.

$$2b = 20$$
$$2b \div 2 = 20 \div 2$$
$$b = 10$$

Therefore, Brittany is 10 years old.

EUREKA
MATH™

Problem Set Sample Solutions

Use tape diagrams to solve each problem.

1. Dwayne scored 55 points in the last basketball game, which is 10 points more than his previous personal best.
    Lebron scored 15 points more than Chris in the same game. Lebron scored the same number of points as Dwayne's
    previous personal best. Let d represent the number of points Dwayne scored during his previous personal best and
    c represent the number of Chris's points.

    a. How many points did Chris score during the game?

    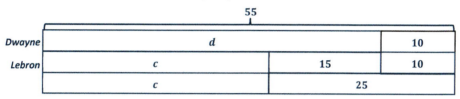

    Equation for Dwayne's Tape Diagram: $d + 10 = 55$

    Equation for Lebron's Tape Diagram:

    $$c + 15 + 10 = 55$$
    $$c + 25 = 55$$
    $$c + 25 - 25 = 55 - 25$$
    $$c = 30$$

    Therefore, Chris scored 30 points in the game.

    b. If these are the only three players who scored, what was the team's total number of points at the end of the
        game?

    Dwayne scored 55 points. Chris scored 30 points. Lebron scored 45 points (answer to Dwayne's equation).
    Therefore, the total number of points scored is $55 + 30 + 45 = 130$.

2. The number of customers at Yummy Smoothies varies throughout the day. During the lunch rush on Saturday, there were 120 customers at Yummy Smoothies. The number of customers at Yummy Smoothies during dinner time was 10 customers fewer than the number during breakfast. The number of customers at Yummy Smoothies during lunch was 3 times more than during breakfast. How many people were at Yummy Smoothies during breakfast? How many people were at Yummy Smoothies during dinner? Let d represent the number of customers at Yummy Smoothies during dinner and b represent the number of customers at Yummy Smoothies during breakfast.

Tape Diagram for Lunch:

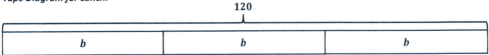

Tape Diagram for Dinner:

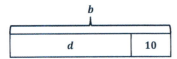

Equation for Lunch's Tape Diagram:

$$3b = 120$$
$$3b \div 3 = 120 \div 3$$
$$b = 40$$

Now that we know 40 customers were at Yummy Smoothies for breakfast, we can use this information and the tape diagram for dinner to determine how many customers were at Yummy Smoothies during dinner.

$$d + 10 = 40$$
$$d + 10 - 10 = 40 - 10$$
$$d = 30$$

Therefore, 30 customers were at Yummy Smoothies during dinner and 40 customers during breakfast.

3. Karter has 24 T-shirts. Karter has 8 fewer pairs of shoes than pairs of pants. If the number of T-shirts Karter has is double the number of pants he has, how many pairs of shoes does Karter have? Let p represent the number of pants Karter has and s represent the number of pairs of shoes he has.

Tape Diagram for T-Shirts:

Tape Diagram for Shoes:

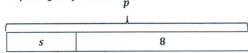

Equation for T-Shirts Tape Diagram:

$$2p = 24$$
$$2p \div 2 = 24 \div 2$$
$$p = 12$$

Equation for Shoes Tape Diagram:

$$s + 8 = 12$$
$$s + 8 - 8 = 12 - 8$$
$$s = 4$$

Karter has 4 pairs of shoes.

4. Darnell completed 35 push-ups in one minute, which is 8 more than his previous personal best. Mia completed 6 more push-ups than Katie. If Mia completed the same amount of push-ups as Darnell completed during his previous personal best, how many push-ups did Katie complete? Let d represent the number of push-ups Darnell completed during his previous personal best and k represent the number of push-ups Katie completed.

35		
d		8
k	6	8
k	14	

$d + 8 = 35$

$$k + 6 + 8 = 35$$
$$k + 14 = 35$$
$$k + 14 - 14 = 35 - 14$$
$$k = 21$$

Katie completed 21 *push-ups.*

5. Justine swims freestyle at a pace of 150 laps per hour. Justine swims breaststroke 20 laps per hour slower than she swims butterfly. If Justine's freestyle speed is three times faster than her butterfly speed, how fast does she swim breaststroke? Let b represent Justine's butterfly speed in laps per hour and r represent Justine's breaststroke speed in laps per hour.

Tape Diagram for Breaststroke:

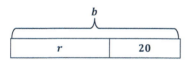

b	
r	20

Tape Diagram for Freestyle:

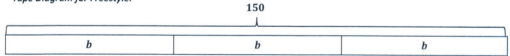

150		
b	b	b

$$3b = 150$$
$$3b \div 3 = 150 \div 3$$
$$b = 50$$

Therefore, Justine swims butterfly at a pace of 50 *laps per hour.*

$$r + 20 = 50$$
$$r + 20 - 20 = 50 - 20$$
$$r = 30$$

Therefore, Justine swims breaststroke at a pace of 30 *laps per hour.*

Number Correct: _____

Addition of Decimals II—Round 1

Directions: Evaluate each expression.

1.	2.5 + 4	
2.	2.5 + 0.4	
3.	2.5 + 0.04	
4.	2.5 + 0.004	
5.	2.5 + 0.0004	
6.	6 + 1.3	
7.	0.6 + 1.3	
8.	0.06 + 1.3	
9.	0.006 + 1.3	
10.	0.0006 + 1.3	
11.	0.6 + 13	
12.	7 + 0.2	
13.	0.7 + 0.02	
14.	0.07 + 0.2	
15.	0.7 + 2	
16.	7 + 0.02	
17.	6 + 0.3	
18.	0.6 + 0.03	
19.	0.06 + 0.3	
20.	0.6 + 3	
21.	6 + 0.03	
22.	0.6 + 0.3	

23.	4.5 + 3.1	
24.	4.5 + 0.31	
25.	4.5 + 0.031	
26.	0.45 + 0.031	
27.	0.045 + 0.031	
28.	12 + 0.36	
29.	1.2 + 3.6	
30.	1.2 + 0.36	
31.	1.2 + 0.036	
32.	0.12 + 0.036	
33.	0.012 + 0.036	
34.	0.7 + 3	
35.	0.7 + 0.3	
36.	0.07 + 0.03	
37.	0.007 + 0.003	
38.	5 + 0.5	
39.	0.5 + 0.5	
40.	0.05 + 0.05	
41.	0.005 + 0.005	
42.	0.11 + 19	
43.	1.1 + 1.9	
44.	0.11 + 0.19	

Lesson 28:   Two-Step Problems—All Operations

EUREKA
MATH™

Addition of Decimals II—Round 1 [KEY]

Directions: Evaluate each expression.

1.	2.5 + 4	6.5	23.	4.5 + 3.1	7.6
2.	2.5 + 0.4	2.9	24.	4.5 + 0.31	4.81
3.	2.5 + 0.04	2.54	25.	4.5 + 0.031	4.531
4.	2.5 + 0.004	2.504	26.	0.45 + 0.031	0.481
5.	2.5 + 0.0004	2.5004	27.	0.045 + 0.031	0.076
6.	6 + 1.3	7.3	28.	12 + 0.36	12.36
7.	0.6 + 1.3	1.9	29.	1.2 + 3.6	4.8
8.	0.06 + 1.3	1.36	30.	1.2 + 0.36	1.56
9.	0.006 + 1.3	1.306	31.	1.2 + 0.036	1.236
10.	0.0006 + 1.3	1.3006	32.	0.12 + 0.036	0.156
11.	0.6 + 13	13.6	33.	0.012 + 0.036	0.048
12.	7 + 0.2	7.2	34.	0.7 + 3	3.7
13.	0.7 + 0.02	0.72	35.	0.7 + 0.3	1
14.	0.07 + 0.2	0.27	36.	0.07 + 0.03	0.1
15.	0.7 + 2	2.7	37.	0.007 + 0.003	0.01
16.	7 + 0.02	7.02	38.	5 + 0.5	5.5
17.	6 + 0.3	6.3	39.	0.5 + 0.5	1
18.	0.6 + 0.03	0.63	40.	0.05 + 0.05	0.1
19.	0.06 + 0.3	0.36	41.	0.005 + 0.005	0.01
20.	0.6 + 3	3.6	42.	0.11 + 19	19.11
21.	6 + 0.03	6.03	43.	1.1 + 1.9	3
22.	0.6 + 0.3	0.9	44.	0.11 + 0.19	0.3

Lesson 28:   Two-Step Problems—All Operations

Number Correct: _____
Improvement: _____

Addition of Decimals II—Round 2

Directions: Evaluate each expression.

1.	7.4 + 3		23.	3.6 + 2.3	
2.	7.4 + 0.3		24.	3.6 + 0.23	
3.	7.4 + 0.03		25.	3.6 + 0.023	
4.	7.4 + 0.003		26.	0.36 + 0.023	
5.	7.4 + 0.0003		27.	0.036 + 0.023	
6.	6 + 2.2		28.	0.13 + 56	
7.	0.6 + 2.2		29.	1.3 + 5.6	
8.	0.06 + 2.2		30.	1.3 + 0.56	
9.	0.006 + 2.2		31.	1.3 + 0.056	
10.	0.0006 + 2.2		32.	0.13 + 0.056	
11.	0.6 + 22		33.	0.013 + 0.056	
12.	7 + 0.8		34.	2 + 0.8	
13.	0.7 + 0.08		35.	0.2 + 0.8	
14.	0.07 + 0.8		36.	0.02 + 0.08	
15.	0.7 + 8		37.	0.002 + 0.008	
16.	7 + 0.08		38.	0.16 + 14	
17.	5 + 0.4		39.	1.6 + 1.4	
18.	0.5 + 0.04		40.	0.16 + 0.14	
19.	0.05 + 0.4		41.	0.016 + 0.014	
20.	0.5 + 4		42.	15 + 0.15	
21.	5 + 0.04		43.	1.5 + 1.5	
22.	5 + 0.4		44.	0.15 + 0.15	

EUREKA
MATH™

Addition of Decimals II—Round 2 [KEY]

Directions: Evaluate each expression.

1.	$7.4 + 3$	**10.4**		23.	$3.6 + 2.3$	**5.9**
2.	$7.4 + 0.3$	**7.7**		24.	$3.6 + 0.23$	**3.83**
3.	$7.4 + 0.03$	**7.43**		25.	$3.6 + 0.023$	**3.623**
4.	$7.4 + 0.003$	**7.403**		26.	$0.36 + 0.023$	**0.383**
5.	$7.4 + 0.0003$	**7.4003**		27.	$0.036 + 0.023$	**0.059**
6.	$6 + 2.2$	**8.2**		28.	$0.13 + 56$	**56.13**
7.	$0.6 + 2.2$	**2.8**		29.	$1.3 + 5.6$	**6.9**
8.	$0.06 + 2.2$	**2.26**		30.	$1.3 + 0.56$	**1.86**
9.	$0.006 + 2.2$	**2.206**		31.	$1.3 + 0.056$	**1.356**
10.	$0.0006 + 2.2$	**2.2006**		32.	$0.13 + 0.056$	**0.186**
11.	$0.6 + 22$	**22.6**		33.	$0.013 + 0.056$	**0.069**
12.	$7 + 0.8$	**7.8**		34.	$2 + 0.8$	**2.8**
13.	$0.7 + 0.08$	**0.78**		35.	$0.2 + 0.8$	**1**
14.	$0.07 + 0.8$	**0.87**		36.	$0.02 + 0.08$	**0.1**
15.	$0.7 + 8$	**8.7**		37.	$0.002 + 0.008$	**0.01**
16.	$7 + 0.08$	**7.08**		38.	$0.16 + 14$	**14.16**
17.	$5 + 0.4$	**5.4**		39.	$1.6 + 1.4$	**3**
18.	$0.5 + 0.04$	**0.54**		40.	$0.16 + 0.14$	**0.3**
19.	$0.05 + 0.4$	**0.45**		41.	$0.016 + 0.014$	**0.03**
20.	$0.5 + 4$	**4.5**		42.	$15 + 0.15$	**15.15**
21.	$5 + 0.04$	**5.04**		43.	$1.5 + 1.5$	**3**
22.	$5 + 0.4$	**5.4**		44.	$0.15 + 0.15$	**0.3**

Lesson 29: Multi-Step Problems—All Operations

Student Outcomes

- Students use their knowledge of simplifying expressions, order of operations, and properties of equality to calculate the solution of multi-step equations. Students use tables to determine their answers.

- Students check to determine if their solutions make the equations true.

Classwork

Example (20 minutes)

Students participate in the discussion by answering the teacher's questions and following along in their student materials.

Example

The school librarian, Mr. Marker, knows the library has $1,400$ books but wants to reorganize how the books are displayed on the shelves. Mr. Marker needs to know how many fiction, nonfiction, and resource books are in the library. He knows that the library has four times as many fiction books as resource books and half as many nonfiction books as fiction books. If these are the only types of books in the library, how many of each type of book are in the library?

 Give students time to work individually or with a partner in order to attempt to make sense of the problem. Students may attempt to solve the problem on their own prior to the following discussion.

Draw a tape diagram to represent the total number of books in the library.

1,400

Draw two more tape diagrams, one to represent the number of fiction books in the library and one to represent the number of resource books in the library.

- Resource Books:

- Fiction Books:

What variable should we use throughout the problem?

We should use r *to represent the number of resource books in the library because it represents the fewest amount of books. Choosing the variable to represent a different type of book would create fractions throughout the problem.*

Write the relationship between resource books and fiction books algebraically.

If we let r *represent the number of resource books, then* $4r$ *represents the number of fiction books.*

Draw a tape diagram to represent the number of nonfiction books.

Nonfiction Books:

 EUREKA MATH™

How did you decide how many sections this tape diagram would have?

There are half as many nonfiction books as fiction books. Since the fiction book tape diagram has four sections, the nonfiction book tape diagram should have two sections.

Represent the number of nonfiction books in the library algebraically.

$2r$ *because that is half as many as fiction books (4r).*

Use the tape diagrams we drew to solve the problem.

We know that combining the tape diagrams for each type of book will leave us with 1,400 total books.

1,400		
r	$4r$	$2r$

Write an equation that represents the tape diagram.

$4r + 2r + r = 1,400$

Determine the value of r.

We can gather like terms and then solve the equation.

$$7r = 1,400$$
$$7r \div 7 = 1,400 \div 7$$
$$r = 200$$

- What does this 200 mean?
  - *There are 200 resource books in the library because r represented the number of resource books.*

How many fiction books are in the library?

There are 800 fiction books in the library because $4(200) = 800$.

How many nonfiction books are in the library?

There are 400 nonfiction books in the library because $2(200) = 400$.

- We can use a different math tool to solve the problem as well. If we were to make a table, how many columns would we need?
  - 4
- Why do we need four columns?
  - *We need to keep track of the number of fiction, nonfiction, and resource books that are in the library, but we also need to keep track of the total number of books.*

Set up a table with four columns, and label each column.

Fiction	Nonfiction	Resource	Total

- Highlight the important information from the word problem that will help us fill out the second row in our table.

  - *The school librarian, Mr. Marker, **knows the library has** $1,400$ **books** but wants to reorganize how the books are displayed on the shelves. Mr. Marker needs to know how many fiction, nonfiction, and resource books are in the library. He knows that the library has **four times as many fiction books as resource books and half as many nonfiction books as fiction books**. If these are the only types of books in the library, how many of each type of book are in the library?*

- Fill out the second row of the table using the algebraic representations.

Fiction	Nonfiction	Resource	Total
$4r$	$2r$	r	$7r$

- If $r = 1$, how many of each type of book would be in the library?

Fiction	Nonfiction	Resource	Total
$4r$	$2r$	r	$7r$
4	2	1	7

- How can we fill out another row of the table?

  - *Substitute different values in for r.*

- Substitute 5 in for r. How many of each type of book would be in the library then?

Fiction	Nonfiction	Resource	Total
4	2	1	7
20	10	5	35

- Does the library have four times as many fiction books as resource books?

  - *Yes, because $5 \cdot 4 = 20$.*

- Does the library have half as many nonfiction books as fiction books?

  - *Yes, because half of 20 is 10.*

- How do we determine how many of each type of book is in the library when there are 1,400 books in the library?

  - *Continue to multiply the rows by the same value, until the total column has 1,400 books.*

At this point, allow students to work individually to determine how many fiction, nonfiction, and resource books are in the library if there are 1,400 total books. Each table may look different because students may choose different values to multiply by. A sample answer is shown below.

Fiction	Nonfiction	Resource	Total
4	2	1	7
20	10	5	35
200	100	50	350
800	400	200	$1,400$

How many fiction books are in the library?

800

EUREKA MATH

> **How many nonfiction books are in the library?**
>
> 400
>
> **How many resource books are in the library?**
>
> 200

■   Let us check and make sure that our answers fit the relationship described in the word problem.

> **Does the library have four times as many fiction books as resource books?**
>
> *Yes, because* $200 \cdot 4 = 800$.
>
> **Does the library have half as many nonfiction books as fiction books?**
>
> *Yes, because half of* 800 *is* 400.
>
> **Does the library have** $1,400$ **books?**
>
> *Yes, because* $800 + 400 + 200 = 1,400$.

Exercises (15 minutes)

Students work in small groups to answer the following problems using tables and algebraic methods.

> **Exercises**
>
> Solve each problem below using tables and algebraic methods. Then, check your answers with the word problems.
>
> 1.   Indiana Ridge Middle School wanted to add a new school sport, so they surveyed the students to determine which sport is most popular. Students were able to choose among soccer, football, lacrosse, or swimming. The same number of students chose lacrosse and swimming. The number of students who chose soccer was double the number of students who chose lacrosse. The number of students who chose football was triple the number of students who chose swimming. If 434 students completed the survey, how many students chose each sport?
>
> | Soccer | Football | Lacrosse | Swimming | Total |
> |--------|----------|----------|----------|-------|
> | 2 | 3 | 1 | 1 | 7 |
>
> *The rest of the table will vary.*
>
> | Soccer | Football | Lacrosse | Swimming | Total |
> |--------|----------|----------|----------|-------|
> | 2 | 3 | 1 | 1 | 7 |
> | 124 | 186 | 62 | 62 | 434 |
>
> 124 *students chose soccer,* 186 *students chose football,* 62 *students chose lacrosse, and* 62 *students chose swimming.*
>
> *We can confirm that these numbers satisfy the conditions of the word problem because lacrosse and swimming were chosen by the same number of students.* 124 *is double* 62, *so soccer was chosen by double the number of students as lacrosse, and* 186 *is triple* 62, *so football was chosen by* 3 *times as many students as swimming. Also,* $124 + 186 + 62 + 62 = 434.$
>
> *Algebraically: Let s represent the number of students who chose swimming. Then,* $2s$ *is the number of students who chose soccer,* $3s$ *is the number of students who chose football, and* s *is the number of students who chose lacrosse.*
>
> $$2s + 3s + s + s = 434$$
> $$7s = 434$$
> $$7s \div 7 = 434 \div 7$$
> $$s = 62$$
>
> *Therefore,* 62 *students chose swimming, and* 62 *students chose lacrosse.* 124 *students chose soccer because* $2(62) = 124$, *and* 186 *students chose football because* $3(62) = 186$.

2. At Prairie Elementary School, students are asked to pick their lunch ahead of time so the kitchen staff will know what to prepare. On Monday, 6 times as many students chose hamburgers as chose salads. The number of students who chose lasagna was one third the number of students who chose hamburgers. If 225 students ordered lunch, how many students chose each option if hamburger, salad, and lasagna were the only three options?

Hamburger	Salad	Lasagna	Total
6	1	2	9

The rest of the table will vary.

Hamburger	Salad	Lasagna	Total
6	1	2	9
150	25	50	225

150 *students chose a hamburger for lunch,* 25 *students chose a salad, and* 50 *students chose lasagna.*

We can confirm that these numbers satisfy the conditions of the word problem because $25 \cdot 6 = 150$, *so hamburgers were chosen by 6 times more students than salads. Also,* $\frac{1}{3} \cdot 150 = 50$, *which means lasagna was chosen by one third of the number of students who chose hamburgers. Finally,* $150 + 25 + 50 = 225$, *which means* 225 *students completed the survey.*

Algebraically: Let s *represent the number of students who chose a salad. Then,* $6s$ *represents the number of students who chose hamburgers, and* $2s$ *represents the number of students who chose lasagna.*

$$6s + s + 2s = 225$$
$$9s = 225$$
$$9s \div 9 = 225 \div 9s$$
$$s = 25$$

This means that 25 *students chose salad,* 150 *students chose hamburgers because* $6(25) = 150$, *and* 50 *students chose lasagna because* $2(25) = 50$.

3. The art teacher, Mr. Gonzalez, is preparing for a project. In order for students to have the correct supplies, Mr. Gonzalez needs 10 times more markers than pieces of construction paper. He needs the same number of bottles of glue as pieces of construction paper. The number of scissors required for the project is half the number of pieces of construction paper. If Mr. Gonzalez collected 400 items for the project, how many of each supply did he collect?

Markers	Construction Paper	Glue Bottles	Scissors	Total
20	2	2	1	25

The rest of the table will vary.

Markers	Construction Paper	Glue Bottles	Scissors	Total
20	2	2	1	25
320	32	32	16	400

Mr. Gonzalez collected 320 *markers,* 32 *pieces of construction paper,* 32 *glue bottles, and* 16 *scissors for the project.*

We can confirm that these numbers satisfy the conditions of the word problem because Mr. Gonzalez collected the same number of pieces of construction paper and glue bottles. Also, $32 \cdot 10 = 320$, *so Mr. Gonzalez collected* 10 *times more markers than pieces of construction paper and glue bottles. Mr. Gonzalez only collected* 16 *pairs of scissors, which is half of the number of pieces of construction paper. The supplies collected add up to* 400 *supplies, which is the number of supplies indicated in the word problem.*

Algebraically: Let s *represent the number of scissors needed for the project, which means* $20s$ *represents the number of markers needed,* $2s$ *represents the number of pieces of construction paper needed, and* $2s$ *represents the number of glue bottles needed.*

$$20s + 2s + 2s + s = 400$$
$$25s = 400$$
$$\frac{25s}{25} = \frac{400}{25}$$
$$s = 16$$

This means that 16 *pairs of scissors,* 320 *markers,* 32 *pieces of construction paper, and* 32 *glue bottles are required for the project.*

EUREKA MATH™

MP.1

4. The math teacher, Ms. Zentz, is buying appropriate math tools to use throughout the year. She is planning on buying twice as many rulers as protractors. The number of calculators Ms. Zentz is planning on buying is one quarter of the number of protractors. If Ms. Zentz buys 65 items, how many protractors does Ms. Zentz buy?

Rulers	Protractors	Calculators	Total
8	4	1	13

The rest of the table will vary.

Rulers	Protractors	Calculators	Total
8	4	1	13
40	20	5	65

Ms. Zentz will buy 20 protractors.

We can confirm that this number satisfies the conditions of the word problem because the number of protractors is half of the number of rulers, and the number of calculators is one fourth of the number of protractors. Also, $40 + 20 + 5 = 65$, so the total matches the total supplies that Ms. Zentz bought.

Algebraically: Let c represent the number of calculators Ms. Zentz needs for the year. Then, $8c$ represents the number of rulers, and $4c$ represents the number of protractors Ms. Zentz will need throughout the year.

$$8c + 4c + c = 65$$
$$13c = 65$$
$$\frac{13c}{13} = \frac{65}{13}$$
$$c = 5$$

Therefore, Ms. Zentz will need 5 calculators, 40 rulers, and 20 protractors throughout the year.

Allow time to answer student questions and discuss answers. In particular, encourage students to compare solution methods with one another, commenting on the accuracy and efficiency of each.

Closing (5 minutes)

- Pam says she only needed two rows in her table to solve each of the problems. How was she able to do this?
  - *Answers will vary. Pam only needed two rows on her table because she found the scale factor from the total in the first row and the total given in the problem. Once this scale factor is determined, it can be used for all the columns in the table because each table is a ratio table.*
- Is there a more efficient way to get to the answer than choosing random values by which to multiply each row?
  - *Find out how many groups of one set of materials it will take to obtain the total amount desired. Then, multiply the entire row by this number.*

Students may need to see a demonstration to fully understand the reasoning. Use the exercises to further explain.

Relate this problem-solving strategy to the ratio tables discussed throughout Module 1.

Exit Ticket (5 minutes)

Lesson 29:   Multi-Step Problems—All Operations

Name _____   Date _____

Lesson 29: Multi-Step Problems—All Operations

Exit Ticket

Solve the problem using tables and equations, and then check your answer with the word problem. Try to find the answer only using two rows of numbers on your table.

A pet store owner, Byron, needs to determine how much food he needs to feed the animals. Byron knows that he needs to order the same amount of bird food as hamster food. He needs four times as much dog food as bird food and needs half the amount of cat food as dog food. If Byron orders 600 packages of animal food, how much dog food does he buy? Let b represent the number of packages of bird food Byron purchased for the pet store.

©2015 Great Minds. eureka-math.org
G6-M4-TE-B4-1.3.1-01.2016

Exit Ticket Sample Solutions

Solve the problem using tables and equations, and then check your answer with the word problem. Try to find the answer only using two rows of numbers on your table.

A pet store owner, Byron, needs to determine how much food he needs to feed the animals. Byron knows that he needs to order the same amount of bird food as hamster food. He needs four times as much dog food as bird food and needs half the amount of cat food as dog food. If Byron orders 600 packages of animal food, how much dog food does he buy? Let b represent the number of packages of bird food Byron purchased for the pet store.

Bird Food	Hamster Food	Dog Food	Cat Food	Total
1	1	4	2	8

The rest of the table will vary (unless they follow suggestions from the Closing).

Bird Food	Hamster Food	Dog Food	Cat Food	Total
1	1	4	2	8
75	75	300	150	600

Byron would need to order 300 packages of dog food.

The answer makes sense because Byron ordered the same amount of bird food and hamster food. The table also shows that Byron ordered four times as much dog food as bird food, and the amount of cat food he ordered is half the amount of dog food. The total amount of pet food Byron ordered was 600 packages, which matches the word problem.

Algebraically: Let b represent the number of packages of bird food Byron purchased for the pet store. Therefore, b also represents the amount of hamster food, $4b$ represents the amount of dog food, and $2b$ represents the amount of cat food required by the pet store.

$$b + b + 4b + 2b = 600$$
$$8b = 600$$
$$8b \div 8 = 600 \div 8$$
$$b = 75$$

Therefore, Byron will order 75 pounds of bird food, which results in 300 pounds of dog food because $4(75) = 300$.

Problem Set Sample Solutions

Create tables to solve the problems, and then check your answers with the word problems.

1. On average, a baby uses three times the number of large diapers as small diapers and double the number of medium diapers as small diapers.

   a. If the average baby uses 2,940 diapers, size large and small, how many of each size would be used?

Small	Medium	Large	Total
3	2	1	6
1,470	980	490	2,940

   An average baby would use 490 small diapers, 980 medium diapers, and 1,470 large diapers.

   The answer makes sense because the number of large diapers is 3 times more than small diapers. The number of medium diapers is double the number of small diapers, and the total number of diapers is 2,940.

Lesson 29: Multi-Step Problems—All Operations

b.   Support your answer with equations.

Let s represent the number of small diapers a baby needs. Therefore, $2s$ represents the number of medium diapers, and $3s$ represents the amount of large diapers a baby needs.

$$s + 2s + 3s = 2,940$$
$$6s = 2,940$$
$$\frac{6s}{6} = \frac{2,940}{6}$$
$$s = 490$$

Therefore, a baby requires 490 small diapers, 980 medium diapers (because $2(490) = 980$), and $1,470$ large diapers (because $3(490) = 1,470$), which matches the answer in part (a).

2.   Tom has three times as many pencils as pens but has a total of 100 writing utensils.

     a.   How many pencils does Tom have?

Pencils	Pens	Total
3	1	4
75	25	100

     b.   How many more pencils than pens does Tom have?

        $75 - 25 = 50$. *Tom has 50 more pencils than pens.*

3.   Serena's mom is planning her birthday party. She bought balloons, plates, and cups. Serena's mom bought twice as many plates as cups. The number of balloons Serena's mom bought was half the number of cups.

     a.   If Serena's mom bought 84 items, how many of each item did she buy?

Balloons	Plates	Cups	Total
1	4	2	7
12	48	24	84

        Serena's mom bought 12 balloons, 48 plates, and 24 cups.

     b.   Tammy brought 12 balloons to the party. How many total balloons were at Serena's birthday party?

        $12 + 12 = 24$. *There were 24 total balloons at the party.*

     c.   If half the plates and all but four cups were used during the party, how many plates and cups were used?

        $\frac{1}{2} \cdot 48 = 24$. *Twenty-four plates were used during the party.*

        $24 - 4 = 20$. *Twenty cups were used during the party.*

4.   Elizabeth has a lot of jewelry. She has four times as many earrings as watches but half the number of necklaces as earrings. Elizabeth has the same number of necklaces as bracelets.

     a.   If Elizabeth has 117 pieces of jewelry, how many earrings does she have?

Earrings	Watches	Necklaces	Bracelets	Total
4	1	2	2	9
52	13	26	26	117

        Elizabeth has 52 earrings, 13 watches, 26 necklaces, and 26 bracelets.

b. **Support your answer with an equation.**

Let w represent the number of watches Elizabeth has. Therefore, 4w represents the number of earrings Elizabeth has, and 2w represents both the number of necklaces and bracelets she has.

$$4w + w + 2w + 2w = 117$$
$$9w = 117$$
$$\frac{9w}{9} = \frac{117}{9}$$
$$w = 13$$

Therefore, Elizabeth has 13 watches, 52 earrings because $4(13) = 52$, and 26 necklaces and bracelets each because $2(13) = 26$.

5. Claudia was cooking breakfast for her entire family. She made double the amount of chocolate chip pancakes as she did regular pancakes. She only made half as many blueberry pancakes as she did regular pancakes. Claudia also knows her family loves sausage, so she made triple the amount of sausage as blueberry pancakes.

    a. How many of each breakfast item did Claudia make if she cooked 90 items in total?

Chocolate Chip Pancakes	Regular Pancakes	Blueberry Pancakes	Sausage	Total
4	2	1	3	10
36	18	9	27	90

Claudia cooked 36 chocolate chip pancakes, 18 regular pancakes, 9 blueberry pancakes, and 27 pieces of sausage.

    b. After everyone ate breakfast, there were 4 chocolate chip pancakes, 5 regular pancakes, 1 blueberry pancake, and no sausage left. How many of each item did the family eat?

The family ate 32 chocolate chip pancakes, 13 regular pancakes, 8 blueberry pancakes, and 27 pieces of sausage during breakfast.

6. During a basketball game, Jeremy scored triple the number of points as Donovan. Kolby scored double the number of points as Donovan.

    a. If the three boys scored 36 points, how many points did each boy score?

Jeremy	Donovan	Kolby	Total
3	1	2	6
18	6	12	36

Jeremy scored 18 points, Donovan scored 6 points, and Kolby scored 12 points.

    b. **Support your answer with an equation.**

Let d represent the number of points Donovan scored, which means 3d represents the number of points Jeremy scored, and 2d represents the number of points Kolby scored.

$$3d + d + 2d = 36$$
$$6d = 36$$
$$\frac{6d}{6} = \frac{36}{6}$$
$$d = 6$$

Therefore, Donovan scored 6 points, Jeremy scored 18 points because $3(6) = 18$, and Kolby scored 12 points because $2(6) = 12$.

Topic H

Applications of Equations

6.EE.B.5, 6.EE.B.6, 6.EE.B.7, 6.EE.B.8, 6.EE.C.9

Focus Standards:	6.EE.B.5	Understand solving an equation or inequality as a process of answering a question: which values from a specified set, if any, make the equation or inequality true? Use substitution to determine whether a given number in a specified set makes an equation or inequality true.
	6.EE.B.6	Use variables to represent numbers and write expressions when solving a real-world or mathematical problem; understand that a variable can represent an unknown number, or, depending on the purpose at hand, any number in a specified set.
	6.EE.B.7	Solve real-world and mathematical problems by writing and solving equations of the form $x + p = q$ and $px = q$ for cases in which p, q, and x are all nonnegative rational numbers.
	6.EE.B.8	Write an inequality of the form $x > c$ or $x < c$ to represent a constraint or condition in a real-world mathematical problem. Recognize that inequalities of the form $x > c$ or $x < c$ have infinitely many solutions; represent solutions of such inequalities on number line diagrams.
	6.EE.C.9	Use variables to represent two quantities in a real-world problem that change in relationship to one another; write an equation to express one quantity, thought of as the dependent variable, in terms of the other quantity, thought of as the independent variable. Analyze the relationship between the dependent and independent variables using graphs and tables, and relate these to the equation. *For example, in a problem involving motion at constant speed, list and graph ordered pairs of distances and times, and write the equation $d = 65t$ to represent the relationship between distance and time.*

EUREKA MATH™

Instructional Days:	5
Lesson 30:	One-Step Problems in the Real World (P)[1]
Lesson 31:	Problems in Mathematical Terms (P)
Lesson 32:	Multi-Step Problems in the Real World (P)
Lesson 33:	From Equations to Inequalities (P)
Lesson 34:	Writing and Graphing Inequalities in Real-World Problems (P)

In Topic H, students apply their knowledge from the entire module to solve equations in real-world, contextual problems. In Lesson 30, students use prior knowledge from Grade 4 to solve missing angle problems. Students write and solve one-step equations in order to determine a missing angle. Lesson 31 involves students using their prior knowledge from Module 1 to construct tables of independent and dependent values in order to analyze equations with two variables from real-life contexts. They represent equations by plotting values from the tables on a coordinate grid in Lesson 32. The module concludes with Lessons 33 and 34, where students refer to true and false number sentences in order to move from solving equations to writing inequalities that represent a constraint or condition in real-life or mathematical problems. Students understand that inequalities have infinitely many solutions and represent those solutions on number line diagrams.

---

[1]Lesson Structure Key: **P**-Problem Set Lesson, **M**-Modeling Cycle Lesson, **E**-Exploration Lesson, **S**-Socratic Lesson

Lesson 30: One-Step Problems in the Real World

Student Outcomes

- Students calculate missing angle measures by writing and solving equations.

Lesson Notes

This is an application lesson based on understandings developed in Grade 4. The three standards applied in this lesson include the following:

4.MD.C.5   Recognize angles as geometric shapes that are formed wherever two rays share a common endpoint, and understand concepts of angle measurement:

   a.   An angle is measured with reference to a circle with its center at the common endpoint of the rays, by considering the fraction of the circular arc between the points where the two rays intersect the circle. An angle that turns through 1/360 of a circle is called a "one-degree angle," and can be used to measure angles.

   b.   An angle that turns through n one-degree angles is said to have an angle measure of n degrees.

4.MD.C.6   Measure angles in whole-number degrees using a protractor. Sketch angles of specified measure.

4.MD.C.7   Recognize angle measure as additive. When an angle is decomposed into non-overlapping parts, the angle measure of the whole is the sum of the angle measures of the parts. Solve addition and subtraction problems to find unknown angles on a diagram in real-world and mathematical problems, e.g., by using an equation with a symbol for the unknown angle measure.

This lesson focuses, in particular, on **4.MD.C.7**.

Classwork

Fluency Exercise (5 minutes): Subtraction of Decimals

Sprint: Refer to Sprints and Sprint Delivery Script sections in the Module Overview for directions on how to administer a Sprint.

Opening Exercise (3 minutes)

Students start the lesson with a review of key angle terms from Grade 4.

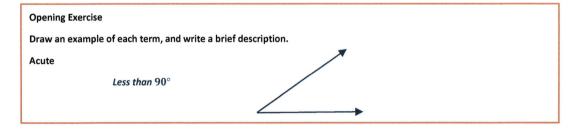

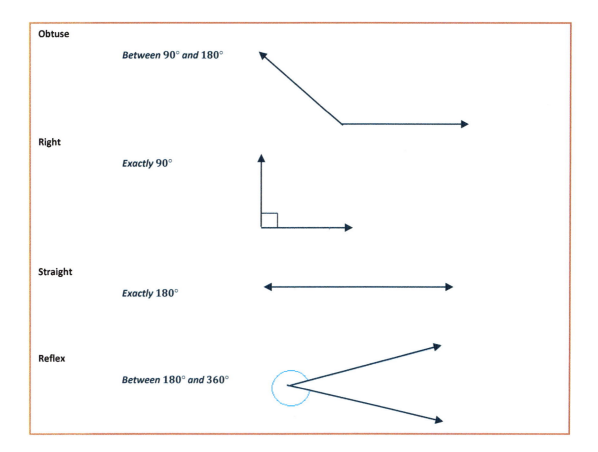

Obtuse

Between 90° and 180°

Right

Exactly 90°

Straight

Exactly 180°

Reflex

Between 180° and 360°

Example 1 (3 minutes)

Example 1

$\angle ABC$ measures 90°. The angle has been separated into two angles. If one angle measures 57°, what is the measure of the other angle?

- In this lesson, we will be using algebra to help us determine unknown measures of angles.

MP.4

How are these two angles related?

The two angles have a sum of 90°.

What equation could we use to solve for x?

$x° + 57° = 90°$

Now, let's solve.

$x° + 57° - 57° = 90° - 57°$

$x° = 33°$

The measure of the unknown angle is 33°.

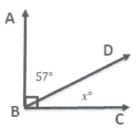

EUREKA
MATH™

Lesson 30:   One-Step Problems in the Real World

333

©2015 Great Minds. eureka-math.org
G6-M4-TE-B4-1.3.1-01.2016

Example 2 (3 minutes)

Example 2

Michelle is designing a parking lot. She has determined that one of the angles should be 115°. What is the measure of angle x and angle y?

How is angle x related to the 115° angle?

The two angles form a straight line. Therefore, they should add up to 180°.

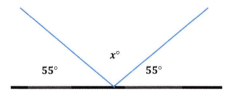

What equation would we use to show this?

$x° + 115° = 180°$

How would you solve this equation?

115° was added to angle x, so I will take away 115° to get back to angle x.

$$x° + 115° - 115° = 180° - 115°$$
$$x° = 65°$$

The angle next to 115°, labeled with an x, is equal to 65°.

How is angle y related to the angle that measures 115°?

These two angles also form a straight line and must add up to 180°.

Therefore, angles x and y must both be equal to 65°.

MP.4

Example 3 (3 minutes)

Example 3

A beam of light is reflected off a mirror. Below is a diagram of the reflected beam. Determine the missing angle measure.

$x°$

$55°$ $55°$

How are the angles in this question related?

There are three angles that, when all placed together, form a straight line. This means that the three angles have a sum of 180°.

What equation could we write to represent the situation?

$55° + x° + 55° = 180°$

MP.4

How would you solve an equation like this?

We can combine the two angles that we do know.

$$55° + 55° + x° = 180°$$
$$110° + x° = 180°$$
$$110° - 110° + x° = 180° - 110°$$
$$x° = 70°$$

The angle of the bounce is $70°$.

Example 4 (3 minutes)

Example 4

$\angle ABC$ measures $90°$. It has been split into two angles, $\angle ABD$ and $\angle DBC$. The measure of $\angle ABD$ and $\angle DBC$ is in a ratio of $4:1$. What are the measures of each angle?

Use a tape diagram to represent the ratio $4:1$.

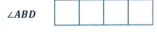

What is the measure of each angle?

5 *units* $= 90°$

1 *unit* $= 18°$

4 *units* $= 72°$

$\angle ABD$ *is* $72°$. $\angle DBC$ *is* $18°$.

How can we represent this situation with an equation?

$$4x° + x° = 90°$$

Solve the equation to determine the measure of each angle.

$$4x° + x° = 90°$$
$$5x° = 90°$$
$$5x° ÷ 5 = 90° ÷ 5$$
$$x° = 18°$$
$$4x° = 4(18°) = 72°$$

The measure of $\angle DBC$ is $18°$ and the measure of $\angle ABD$ is $72°$.

Lesson 30:   One-Step Problems in the Real World

Exercises (15 minutes)

Students work independently.

Exercises

Write and solve an equation in each of the problems.

1. $\angle ABC$ measures $90°$. It has been split into two angles, $\angle ABD$ and $\angle DBC$. The measure of the two angles is in a ratio of $2:1$. What are the measures of each angle?

$$x° + 2x° = 90^0$$
$$3x° = 90°$$
$$\frac{3x°}{3} = \frac{90°}{3}$$
$$x° = 30°$$

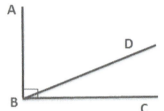

One of the angles measures $30°$, and the other measures $60°$.

2. Solve for x.

$$x° + 64° + 37° = 180°$$
$$x° + 101° = 180°$$
$$x° + 101° - 101° = 180° - 101°$$
$$x° = 79°$$

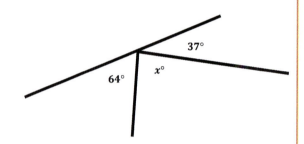

3. Candice is building a rectangular piece of a fence according to the plans her boss gave her. One of the angles is not labeled. Write an equation, and use it to determine the measure of the unknown angle.

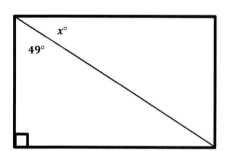

$$x° + 49° = 90°$$
$$x° + 49° - 49° = 90° - 49°$$
$$x° = 41°$$

©2015 Great Minds. eureka-math.org
G6-M4-TE-B4-1.3.1-01.2016

EUREKA
MATH™

4. Rashid hit a hockey puck against the wall at a 38° angle. The puck hit the wall and traveled in a new direction. Determine the missing angle in the diagram.

$$38° + x° + 38° = 180°$$
$$76° + x° = 180°$$
$$76° - 76° + x° = 180° - 76°$$
$$x° = 104°$$

The measure of the missing angle is 104°.

5. Jaxon is creating a mosaic design on a rectangular table. He has added two pieces to one of the corners. The first piece has an angle measuring 38° and is placed in the corner. A second piece has an angle measuring 27° and is also placed in the corner. Draw a diagram to model the situation. Then, write an equation, and use it to determine the measure of the unknown angle in a third piece that could be added to the corner of the table.

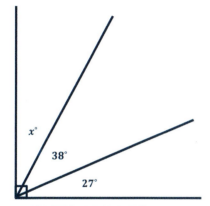

$$x° + 38° + 27° = 90°$$
$$x° + 65° = 90°$$
$$x° + 65° - 65° = 90° - 65°$$
$$x° = 25°$$

The measure of the unknown angle is 25°.

Closing (3 minutes)

- Explain how you determined the equation you used to solve for the missing angle or variable.

  □ *I used the descriptions in the word problems. For example, if it said "the sum of the angles," I knew to add the measures together.*

  □ *I also used my knowledge of angles to know the total angle measure. For example, I know a straight angle has a measure of 180°, and a right angle or a corner has a measure of 90°.*

Exit Ticket (7 minutes)

Name _____   Date _____

Lesson 30: One-Step Problems in the Real World

Write an equation, and solve for the missing angle in each question.

1. Alejandro is repairing a stained glass window. He needs to take it apart to repair it. Before taking it apart, he makes a sketch with angle measures to put it back together.

    Write an equation, and use it to determine the measure of the unknown angle.

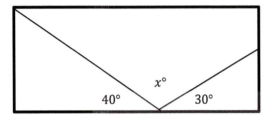

2. Hannah is putting in a tile floor. She needs to determine the angles that should be cut in the tiles to fit in the corner. The angle in the corner measures 90°. One piece of the tile will have a measure of 38°. Write an equation, and use it to determine the measure of the unknown angle.

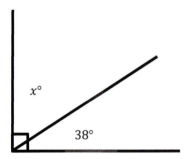

EUREKA
MATH™

Exit Ticket Sample Solutions

Write an equation, and solve for the missing angle in each question.

1. Alejandro is repairing a stained glass window. He needs to take it apart to repair it. Before taking it apart, he makes a sketch with angle measures to put it back together.

   Write an equation, and use it to determine the measure of the unknown angle.

   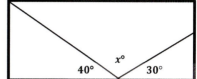

   $$40° + x° + 30° = 180°$$
   $$x° + 40° + 30° = 180°$$
   $$x° + 70° = 180°$$
   $$x° + 70° - 70° = 180° - 70°$$
   $$x° = 110°$$

   The missing angle measures $110°$.

2. Hannah is putting in a tile floor. She needs to determine the angles that should be cut in the tiles to fit in the corner. The angle in the corner measures $90°$. One piece of the tile will have a measure of $38°$. Write an equation, and use it to determine the measure of the unknown angle.

   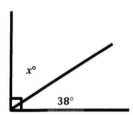

   $$x° + 38° = 90°$$
   $$x° + 38° - 38° = 90° - 38°$$
   $$x° = 52°$$

   The measure of the unknown angle is $52°$.

Problem Set Sample Solutions

Write and solve an equation for each problem.

1. Solve for x.

   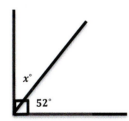

   $$x° + 52° = 90°$$
   $$x° + 52° - 52° = 90° - 52°$$
   $$x° = 38°$$

   The measure of the missing angle is $38°$.

2. $\angle BAE$ measures $90°$. Solve for x.

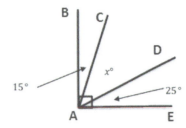

$$15° + x° + 25° = 90°$$
$$15° + 25° + x° = 90°$$
$$40° + x° = 90°$$
$$40° - 40° + x° = 90° - 40°$$
$$x° = 50°$$

3. Thomas is putting in a tile floor. He needs to determine the angles that should be cut in the tiles to fit in the corner. The angle in the corner measures $90°$. One piece of the tile will have a measure of $24°$. Write an equation, and use it to determine the measure of the unknown angle.

$$x° + 24° = 90°$$
$$x° + 24° - 24° = 90° - 24°$$
$$x° = 66°$$

The measure of the unknown angle is $66°$.

4. Solve for x.

$$x° + 105° + 62° = 180°$$
$$x° + 167° = 180°$$
$$x° + 167° - 167° = 180° - 167°$$
$$x° = 13°$$

The measure of the missing angle is $13°$.

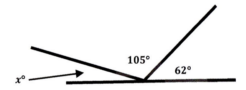

5. Aram has been studying the mathematics behind pinball machines. He made the following diagram of one of his observations. Determine the measure of the missing angle.

$$52° + x° + 68° = 180°$$
$$120° + x° = 180°$$
$$120° + x° - 120° = 180° - 120°$$
$$x° = 60°$$

The measure of the missing angle is $60°$.

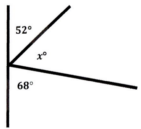

6. The measures of two angles have a sum of $90°$. The measures of the angles are in a ratio of $2:1$. Determine the measures of both angles.

$$2x° + x° = 90°$$
$$3x° = 90°$$
$$\frac{3x°}{3} = \frac{90°}{3}$$
$$x° = 30°$$

The angles measure $30°$ and $60°$.

7. The measures of two angles have a sum of $180°$. The measures of the angles are in a ratio of $5:1$. Determine the measures of both angles.

$$5x° + x° = 180°$$
$$6x° = 180°$$
$$\frac{6x°}{6} = \frac{180}{6}°$$
$$x° = 30°$$

The angles measure $30°$ and $150°$.

Number Correct: _____

Subtraction of Decimals—Round 1

Directions: Evaluate each expression.

1.	$55 - 50$		23.	$9.9 - 5$	
2.	$55 - 5$		24.	$9.9 - 0.5$	
3.	$5.5 - 5$		25.	$0.99 - 0.5$	
4.	$5.5 - 0.5$		26.	$0.99 - 0.05$	
5.	$88 - 80$		27.	$4.7 - 2$	
6.	$88 - 8$		28.	$4.7 - 0.2$	
7.	$8.8 - 8$		29.	$0.47 - 0.2$	
8.	$8.8 - 0.8$		30.	$0.47 - 0.02$	
9.	$33 - 30$		31.	$8.4 - 1$	
10.	$33 - 3$		32.	$8.4 - 0.1$	
11.	$3.3 - 3$		33.	$0.84 - 0.1$	
12.	$1 - 0.3$		34.	$7.2 - 5$	
13.	$1 - 0.03$		35.	$7.2 - 0.5$	
14.	$1 - 0.003$		36.	$0.72 - 0.5$	
15.	$0.1 - 0.03$		37.	$0.72 - 0.05$	
16.	$4 - 0.8$		38.	$8.6 - 7$	
17.	$4 - 0.08$		39.	$8.6 - 0.7$	
18.	$4 - 0.008$		40.	$0.86 - 0.7$	
19.	$0.4 - 0.08$		41.	$0.86 - 0.07$	
20.	$9 - 0.4$		42.	$5.1 - 4$	
21.	$9 - 0.04$		43.	$5.1 - 0.4$	
22.	$9 - 0.004$		44.	$0.51 - 0.4$	

Lesson 30:   One-Step Problems in the Real World

EUREKA MATH

Subtraction of Decimals—Round 1 [KEY]

Directions: Evaluate each expression.

1.	$55 - 50$	**5**	23.	$9.9 - 5$	**4.9**
2.	$55 - 5$	**50**	24.	$9.9 - 0.5$	**9.4**
3.	$5.5 - 5$	**0.5**	25.	$0.99 - 0.5$	**0.49**
4.	$5.5 - 0.5$	**5**	26.	$0.99 - 0.05$	**0.94**
5.	$88 - 80$	**8**	27.	$4.7 - 2$	**2.7**
6.	$88 - 8$	**80**	28.	$4.7 - 0.2$	**4.5**
7.	$8.8 - 8$	**0.8**	29.	$0.47 - 0.2$	**0.27**
8.	$8.8 - 0.8$	**8**	30.	$0.47 - 0.02$	**0.45**
9.	$33 - 30$	**3**	31.	$8.4 - 1$	**7.4**
10.	$33 - 3$	**30**	32.	$8.4 - 0.1$	**8.3**
11.	$3.3 - 3$	**0.3**	33.	$0.84 - 0.1$	**0.74**
12.	$1 - 0.3$	**0.7**	34.	$7.2 - 5$	**2.2**
13.	$1 - 0.03$	**0.97**	35.	$7.2 - 0.5$	**6.7**
14.	$1 - 0.003$	**0.997**	36.	$0.72 - 0.5$	**0.22**
15.	$0.1 - 0.03$	**0.07**	37.	$0.72 - 0.05$	**0.67**
16.	$4 - 0.8$	**3.2**	38.	$8.6 - 7$	**1.6**
17.	$4 - 0.08$	**3.92**	39.	$8.6 - 0.7$	**7.9**
18.	$4 - 0.008$	**3.992**	40.	$0.86 - 0.7$	**0.16**
19.	$0.4 - 0.08$	**0.32**	41.	$0.86 - 0.07$	**0.79**
20.	$9 - 0.4$	**8.6**	42.	$5.1 - 4$	**1.1**
21.	$9 - 0.04$	**8.96**	43.	$5.1 - 0.4$	**4.7**
22.	$9 - 0.004$	**8.996**	44.	$0.51 - 0.4$	**0.11**

Number Correct: _____
Improvement: _____

Subtraction of Decimals—Round 2

Directions: Evaluate each expression.

1.	$66 - 60$	
2.	$66 - 6$	
3.	$6.6 - 6$	
4.	$6.6 - 0.6$	
5.	$99 - 90$	
6.	$99 - 9$	
7.	$9.9 - 9$	
8.	$9.9 - 0.9$	
9.	$22 - 20$	
10.	$22 - 2$	
11.	$2.2 - 2$	
12.	$3 - 0.4$	
13.	$3 - 0.04$	
14.	$3 - 0.004$	
15.	$0.3 - 0.04$	
16.	$8 - 0.2$	
17.	$8 - 0.02$	
18.	$8 - 0.002$	
19.	$0.8 - 0.02$	
20.	$5 - 0.1$	
21.	$5 - 0.01$	
22.	$5 - 0.001$	

23.	$6.8 - 4$	
24.	$6.8 - 0.4$	
25.	$0.68 - 0.4$	
26.	$0.68 - 0.04$	
27.	$7.3 - 1$	
28.	$7.3 - 0.1$	
29.	$0.73 - 0.1$	
30.	$0.73 - 0.01$	
31.	$9.5 - 2$	
32.	$9.5 - 0.2$	
33.	$0.95 - 0.2$	
34.	$8.3 - 5$	
35.	$8.3 - 0.5$	
36.	$0.83 - 0.5$	
37.	$0.83 - 0.05$	
38.	$7.2 - 4$	
39.	$7.2 - 0.4$	
40.	$0.72 - 0.4$	
41.	$0.72 - 0.04$	
42.	$9.3 - 7$	
43.	$9.3 - 0.7$	
44.	$0.93 - 0.7$	

EUREKA MATH

Subtraction of Decimals—Round 2 [KEY]

Directions: Evaluate each expression.

1.	$66 - 60$	**6**	23.	$6.8 - 4$	**2.8**
2.	$66 - 6$	**60**	24.	$6.8 - 0.4$	**6.4**
3.	$6.6 - 6$	**0.6**	25.	$0.68 - 0.4$	**0.28**
4.	$6.6 - 0.6$	**6**	26.	$0.68 - 0.04$	**0.64**
5.	$99 - 90$	**9**	27.	$7.3 - 1$	**6.3**
6.	$99 - 9$	**90**	28.	$7.3 - 0.1$	**7.2**
7.	$9.9 - 9$	**0.9**	29.	$0.73 - 0.1$	**0.63**
8.	$9.9 - 0.9$	**9**	30.	$0.73 - 0.01$	**0.72**
9.	$22 - 20$	**2**	31.	$9.5 - 2$	**7.5**
10.	$22 - 2$	**20**	32.	$9.5 - 0.2$	**9.3**
11.	$2.2 - 2$	**0.2**	33.	$0.95 - 0.2$	**0.75**
12.	$3 - 0.4$	**2.6**	34.	$8.3 - 5$	**3.3**
13.	$3 - 0.04$	**2.96**	35.	$8.3 - 0.5$	**7.8**
14.	$3 - 0.004$	**2.996**	36.	$0.83 - 0.5$	**0.33**
15.	$0.3 - 0.04$	**0.26**	37.	$0.83 - 0.05$	**0.78**
16.	$8 - 0.2$	**7.8**	38.	$7.2 - 4$	**3.2**
17.	$8 - 0.02$	**7.98**	39.	$7.2 - 0.4$	**6.8**
18.	$8 - 0.002$	**7.998**	40.	$0.72 - 0.4$	**0.32**
19.	$0.8 - 0.02$	**0.78**	41.	$0.72 - 0.04$	**0.68**
20.	$5 - 0.1$	**4.9**	42.	$9.3 - 7$	**2.3**
21.	$5 - 0.01$	**4.99**	43.	$9.3 - 0.7$	**8.6**
22.	$5 - 0.001$	**4.999**	44.	$0.93 - 0.7$	**0.23**

©2015 Great Minds. eureka-math.org
G6-M4-TE-B4-1.3.1-01.2016

Lesson 31: Problems in Mathematical Terms

Student Outcomes

- Students analyze an equation in two variables to choose an independent variable and a dependent variable. Students determine whether or not the equation is solved for the second variable in terms of the first variable or vice versa. They then use this information to determine which variable is the independent variable and which is the dependent variable.

- Students create a table by placing the independent variable in the first row or column and the dependent variable in the second row or column. They compute entries in the table by choosing arbitrary values for the independent variable (no constraints) and then determine what the dependent variable must be.

Classwork

Example 1 (10 minutes)

> **Example 1**
>
> Marcus reads for 30 minutes each night. He wants to determine the total number of minutes he will read over the course of a month. He wrote the equation $t = 30d$ to represent the total amount of time that he has spent reading, where t represents the total number of minutes read and d represents the number of days that he read during the month. Determine which variable is independent and which is dependent. Then, create a table to show how many minutes he has read in the first seven days.
>
> | Number of Days (d) | Total Minutes Read (30d) |
> |:---:|:---:|
> | 1 | 30 |
> | 2 | 60 |
> | 3 | 90 |
> | 4 | 120 |
> | 5 | 150 |
> | 6 | 180 |
> | 7 | 210 |
>
> Independent variable   _Number of days_
>
> Dependent variable   _Total minutes read_

MP.1

- When setting up a table, we want the independent variable in the first column and the dependent variable in the second column.

- What do independent and dependent mean?
  - *The independent variable changes, and when it does, it affects the dependent variable. So, the dependent variable depends on the independent variable.*

- In this example, which would be the independent variable, and which would be the dependent variable?
  - *The dependent variable is the total number of minutes read because it depends on how many days Marcus reads. The independent variable is the number of days that Marcus reads.*

- How could you use the table of values to determine the equation if it had not been given?
  - *The number of minutes read shown in the table is always 30 times the number of days. So, the equation would need to show that the total number of minutes read is equal to the number of days times 30.*

Example 2 (5 minutes)

> **Example 2**
>
> Kira designs websites. She can create three different websites each week. Kira wants to create an equation that will give her the total number of websites she can design given the number of weeks she works. Determine the independent and dependent variables. Create a table to show the number of websites she can design over the first 5 weeks. Finally, write an equation to represent the number of websites she can design when given any number of weeks.
>
> Independent variable____# of weeks worked_____
>
> Dependent variable____# of websites designed_____
>
> | # of Weeks Worked (w) | # of Websites Designed (d) |
> | --- | --- |
> | 1 | 3 |
> | 2 | 6 |
> | 3 | 9 |
> | 4 | 12 |
> | 5 | 15 |
>
> Equation____$d = 3w$, where w is the number of weeks worked and d is the number of websites designed.____

- How did you determine which is the dependent variable and which is the independent variable?
  - *Because the number of websites she can make depends on how many weeks she works, I determined that the number of weeks worked was the independent variable, and the number of websites designed was the dependent variable.*

- Does knowing which one is independent and which one is dependent help you write the equation?
  - *I can write the equation and solve for the dependent variable by knowing how the independent variable will affect the dependent variable. In this case, I knew that every week 3 more websites could be completed, so then I multiplied the number of weeks by 3.*

MP.1

Example 3 (5 minutes)

> **Example 3**
>
> Priya streams movies through a company that charges her a $5 monthly fee plus $1.50 per movie. Determine the independent and dependent variables, write an equation to model the situation, and create a table to show the total cost per month given that she might stream between 4 and 10 movies in a month.
>
> Independent variable____# of movies watched per month____
>
> Dependent variable____Total cost per month, in dollars____
>
> Equation____$c = 1.5m + 5$ or $c = 1.50m + 5$____
>
> | # of Movies (m) | Total Cost Per Month, in dollars (c) |
> | --- | --- |
> | 4 | 11 |
> | 5 | 12.50 |
> | 6 | 14 |
> | 7 | 15.50 |
> | 8 | 17 |
> | 9 | 18.50 |
> | 10 | 20 |

- Is the flat fee an independent variable, a dependent variable, or neither?
  - *The $5 flat fee is neither. It is not causing the change in the dependent value, and it is not changing. Instead, the $5 flat fee is a constant that is added on each month.*

- Why isn't the equation $c = 5m + 1.50$?

  □   *The $5 fee is only paid once a month. m is the number of movies watched per month, so it needs to be multiplied by the price per movie, which is $1.50.*

Exercises (15 minutes)

Students work in pairs or independently.

Exercises

1.   Sarah is purchasing pencils to share. Each package has 12 pencils. The equation $n = 12p$, where n is the total number of pencils and p is the number of packages, can be used to determine the total number of pencils Sarah purchased. Determine which variable is dependent and which is independent. Then, make a table showing the number of pencils purchased for 3–7 packages.

The number of packages, p, is the independent variable.

The total number of pencils, n, is the dependent variable.

# of Packages (p)	Total # of Pencils (n = 12p)
3	36
4	48
5	60
6	72
7	84

2.   Charlotte reads 4 books each week. Let b be the number of books she reads each week, and let w be the number of weeks that she reads. Determine which variable is dependent and which is independent. Then, write an equation to model the situation, and make a table that shows the number of books read in under 6 weeks.

The number of weeks, w, is the independent variable.

The number of books, b, is the dependent variable.

$b = 4w$

# of Weeks (w)	# of Books (b = 4w)
1	4
2	8
3	12
4	16
5	20

Lesson 31:   Problems in Mathematical Terms

©2015 Great Minds. eureka-math.org
G6-M4-TE-B4-1.3.1-01.2016

3. A miniature golf course has a special group rate. You can pay $20 plus $3 per person when you have a group of 5 or more friends. Let f be the number of friends and c be the total cost. Determine which variable is independent and which is dependent, and write an equation that models the situation. Then, make a table to show the cost for 5 to 12 friends.

The number of friends, f, is the independent variable.

The total cost in dollars, c, is the dependent variable.

$c = 3f + 20$

# of Friends (f)	Total Cost, in dollars ($c = 3f + 20$)
5	35
6	38
7	41
8	44
9	47
10	50
11	53
12	56

4. Carlos is shopping for school supplies. He bought a pencil box for $3, and he also needs to buy notebooks. Each notebook is $2. Let t represent the total cost of the supplies and n be the number of notebooks Carlos buys. Determine which variable is independent and which is dependent, and write an equation that models the situation. Then, make a table to show the cost for 1 to 5 notebooks.

The total number of notebooks, n, is the independent variable.

The total cost in dollars, t, is the dependent variable.

$$t = 2n + 3$$

# of Notebooks (n)	Total Cost, in dollars ($t = 2n + 3$)
1	5
2	7
3	9
4	11
5	13

Closing (5 minutes)

Use this time for partners to share their answers from the exercises with another set of partners.

- How can you determine which variable is independent and which variable is dependent?
    - *The dependent variable is affected by changes in the independent variable.*
    - *I can write a sentence stating that one variable depends on another. For example, the amount of money earned depends on the number of hours worked. So, the money earned is the dependent variable, and the number of hours worked is the independent variable.*

Exit Ticket (5 minutes)

©2015 Great Minds. eureka-math.org
G6-M4-TE-B4-1.3.1-01.2016

Name _____   Date _____

Lesson 31: Problems in Mathematical Terms

For each problem, determine the independent and dependent variables, write an equation to represent the situation, and then make a table with at least 5 values that models the situation.

1. Kyla spends 60 minutes of each day exercising. Let d be the number of days that Kyla exercises, and let m represent the total minutes of exercise in a given time frame. Show the relationship between the number of days that Kyla exercises and the total minutes that she exercises.

Independent variable _____

Dependent variable _____

Equation _____

2. A taxicab service charges a flat fee of $8 plus an additional $1.50 per mile. Show the relationship between the total cost and the number of miles driven.

Independent variable _____

Dependent variable _____

Equation _____

Exit Ticket Sample Solutions

For each problem, determine the independent and dependent variables, write an equation to represent the situation, and then make a table with at least 5 values that models the situation.

1. Kyla spends 60 minutes of each day exercising. Let d be the number of days that Kyla exercises, and let m represent the total minutes of exercise in a given time frame. Show the relationship between the number of days that Kyla exercises and the total minutes that she exercises.

    Tables may vary.

# of Days	# of Minutes
0	0
1	60
2	120
3	180
4	240

    Independent variable ____*Number of Days*____

    Dependent variable ____*Total Number of Minutes*____

    Equation ____$m = 60d$____

2. A taxicab service charges a flat fee of $8 plus an additional $1.50 per mile. Show the relationship between the total cost and the number of miles driven.

    Tables may vary.

# of Miles	Total Cost, in dollars
0	8.00
1	9.50
2	11.00
3	12.50
4	14.00

    Independent variable ____*Number of miles*____

    Dependent variable ____*Total cost, in dollars*____

    Equation ____$c = 1.50m + 8$____

Problem Set Sample Solutions

1. Jaziyah sells 3 houses each month. To determine the number of houses she can sell in any given number of months, she uses the equation $t = 3m$, where t is the total number of houses sold and m is the number of months. Name the independent and dependent variables. Then, create a table to show how many houses she sells in fewer than 6 months.

    The independent variable is the number of months. The dependent variable is the total number of houses sold.

# of Months	Total Number of Houses
1	3
2	6
3	9
4	12
5	15

Lesson 31:   Problems in Mathematical Terms

2.   Joshua spends 25 minutes of each day reading. Let d be the number of days that he reads, and let m represent the total minutes of reading. Determine which variable is independent and which is dependent. Then, write an equation that models the situation. Make a table showing the number of minutes spent reading over 7 days.

The number of days, d, is the independent variable.

The total number of minutes of reading, m, is the dependent variable.

$m = 25d$

# of Days	# of Minutes
1	25
2	50
3	75
4	100
5	125
6	150
7	175

3.   Each package of hot dog buns contains 8 buns. Let p be the number of packages of hot dog buns and b be the total number of buns. Determine which variable is independent and which is dependent. Then, write an equation that models the situation, and make a table showing the number of hot dog buns in 3 to 8 packages.

The number of packages, p, is the independent variable.

The total number of hot dog buns, b, is the dependent variable.

$b = 8p$

# of Packages	Total # of Hot Dog Buns
3	24
4	32
5	40
6	48
7	56
8	64

4.   Emma was given 5 seashells. Each week she collected 3 more. Let w be the number of weeks Emma collects seashells and s be the number of seashells she has total. Which variable is independent, and which is dependent? Write an equation to model the relationship, and make a table to show how many seashells she has from week 4 to week 10.

The number of weeks, w, is the independent variable.

The total number of seashells, s, is the dependent variable.

$s = 3w + 5$

# of Weeks	Total # of Seashells
4	17
5	20
6	23
7	26
8	29
9	32
10	35

©2015 Great Minds. eureka-math.org
G6-M4-TE-B4-1.3.1-01.2016

5. Emilia is shopping for fresh produce at a farmers' market. She bought a watermelon for $5, and she also wants to buy peppers. Each pepper is $0.75. Let t represent the total cost of the produce and n be the number of peppers bought. Determine which variable is independent and which is dependent, and write an equation that models the situation. Then, make a table to show the cost for 1 to 5 peppers.

The number of peppers, n, is the independent variable.

The total cost in dollars, t, is the dependent variable.

$t = 0.75n + 5$

# of Peppers	Total Cost, in dollars
1	5.75
2	6.50
3	7.25
4	8.00
5	8.75

6. A taxicab service charges a flat fee of $7 plus an additional $1.25 per mile driven. Show the relationship between the total cost and the number of miles driven. Which variable is independent, and which is dependent? Write an equation to model the relationship, and make a table to show the cost of 4 to 10 miles.

The number of miles driven, m, is the independent variable.

The total cost in dollars, c, is the dependent variable.

$c = 1.25m + 7$

# of Miles	Total Cost, in dollars
4	12.00
5	13.25
6	14.50
7	15.75
8	17.00
9	18.25
10	19.50

Lesson 32: Multi-Step Problems in the Real World

Student Outcomes

- Students analyze an equation in two variables, choose an independent variable and a dependent variable, make a table, and make a graph for the equation by plotting the points in the table. For the graph, the independent variable is usually represented by the horizontal axis, and the dependent variable is usually represented by the vertical axis.

Classwork

Opening Exercise (5 minutes)

Opening Exercise

Xin is buying beverages for a party that come in packs of 8. Let p be the number of packages Xin buys and t be the total number of beverages. The equation $t = 8p$ can be used to calculate the total number of beverages when the number of packages is known. Determine the independent and dependent variables in this scenario. Then, make a table using whole number values of p less than 6.

The total number of beverages is the dependent variable because the total number of beverages depends on the number of packages purchased. Therefore, the independent variable is the number of packages purchased.

Number of Packages (p)	Total Number of Beverages ($t = 8p$)
0	0
1	8
2	16
3	24
4	32
5	40

Example 1 (7 minutes)

Example 1

Make a graph for the table in the Opening Exercise.

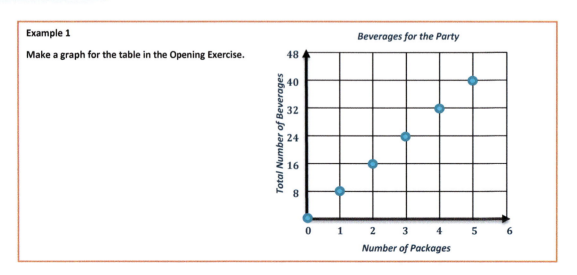

Beverages for the Party

(vertical axis: Total Number of Beverages — 8, 16, 24, 32, 40, 48)
(horizontal axis: Number of Packages — 0, 1, 2, 3, 4, 5, 6)

- To make a graph, we must determine which variable is measured along the horizontal axis and which variable is measured along the vertical axis.
- Generally, the independent variable is measured along the x-axis. Which axis is the x-axis?
  - *The x-axis is the horizontal axis.*
- Where would you put the dependent variable?
  - *On the y-axis. It travels vertically, or up and down.*
- We want to show how the number of beverages changes when the number of packages changes. To check that you have set up your graph correctly, try making a sentence out of the labels on the axes. Write your sentence using the label from the y-axis first followed by the label from the x-axis. The total number of beverages depends on the number of packages purchased.

Example 2 (3 minutes)

MP.2

Example 2

Use the graph to determine which variable is the independent variable and which is the dependent variable. Then, state the relationship between the quantities represented by the variables.

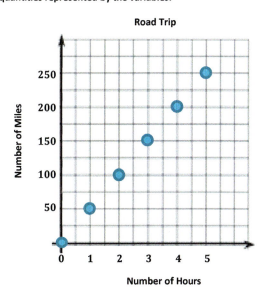

The number of miles driven depends on how many hours they drive. Therefore, the number of miles driven is the dependent variable, and the number of hours is the independent variable. This graph shows that they can travel 50 miles every hour. So, the total number of miles driven increases by 50 every time the number of hours increases by 1.

Exercise (20 minutes)

Students work individually.

Exercises

1. Each week Quentin earns $30. If he saves this money, create a graph that shows the total amount of money Quentin has saved from week 1 through week 8. Write an equation that represents the relationship between the number of weeks that Quentin has saved his money, w, and the total amount of money in dollars he has saved, s. Then, name the independent and dependent variables. Write a sentence that shows this relationship.

$s = 30w$

The amount of money saved in dollars, s, is the dependent variable, and the number of weeks, w, is the independent variable.

Number of Weeks	Total Saved ($)
1	30
2	60
3	90
4	120
5	150
6	180
7	210
8	240

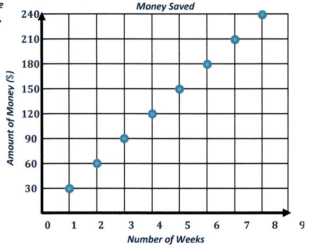

Therefore, the amount of money Quentin has saved increases by $30 for every week he saves money.

2. Zoe is collecting books to donate. She started with 3 books and collects two more each week. She is using the equation $b = 2w + 3$, where b is the total number of books collected and w is the number of weeks she has been collecting books. Name the independent and dependent variables. Then, create a graph to represent how many books Zoe has collected when w is 5 or less.

The number of weeks is the independent variable. The number of books collected is the dependent variable.

Number of Weeks	Number of Books Collected
0	3
1	5
2	7
3	9
4	11
5	13

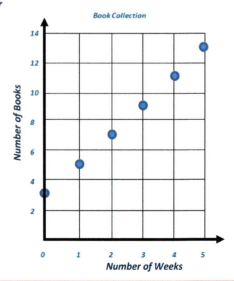

EUREKA MATH™

3. Eliana plans to visit the fair. She must pay $5 to enter the fairgrounds and an additional $3 per ride. Write an equation to show the relationship between r, the number of rides, and t, the total cost in dollars. State which variable is dependent and which is independent. Then, create a graph that models the equation.

$t = 3r + 5$

The number of rides is the independent variable, and the total cost in dollars, is the dependent variable.

# of Rides	Total Cost (in dollars)
0	5
1	8
2	11
3	14
4	17

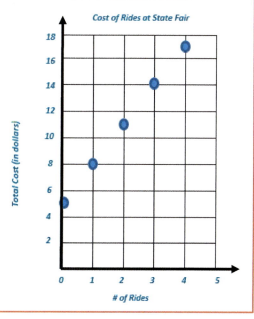

Closing (5 minutes)

▪ Imagine that you are helping a friend with his math work. Here is the problem he was solving:

Henry is taking a taxicab home. The cab company charges an initial fee of $5 plus $2 for each additional mile. Henry uses the equation $t = 2m + 5$ to calculate the cost of the ride, where t is the total cost and m is the number of miles.

Your friend states that t is the dependent variable and m is the independent variable. Then, the friend starts to sketch a graph.

▪ What would you tell your friend when looking over her work?

  □ *I would tell my friend that the dependent variable should go on the vertical axis or the y-axis. Then your graph will show that the total cost of the ride depends on how many miles you travel in the taxicab.*

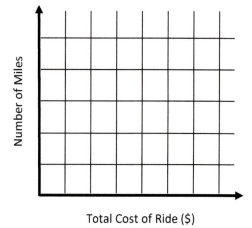

Exit Ticket (5 minutes)

Name _____ Date _____

Lesson 32: Multi-Step Problems in the Real World

Exit Ticket

Determine which variable is the independent variable and which variable is the dependent variable. Write an equation, make a table, and plot the points from the table on the graph.

Enoch can type 40 words per minute. Let w be the number of words typed and m be the number of minutes spent typing.

Independent variable _____

Dependent variable _____

Equation _____

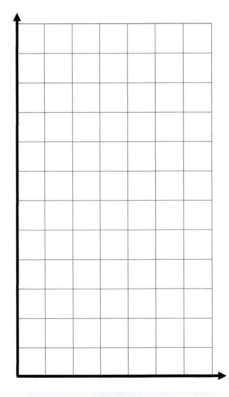

EUREKA MATH™

Exit Ticket Sample Solutions

Determine which variable is the independent variable and which variable is the dependent variable. Write an equation, make a table, and plot the points from the table on the graph.

Enoch can type 40 words per minutes. Let w be the number of words typed and m be the number of minutes spent typing.

The independent variable is the number of minutes spent typing. The dependent variable is the number of words typed.

The equation is $w = 40m$.

# of Minutes	# of Words
0	0
1	40
2	80
3	120
4	160
5	200

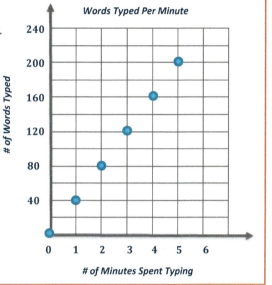

Problem Set Sample Solutions

1. Caleb started saving money in a cookie jar. He started with $\$25$. He adds $\$10$ to the cookie jar each week. Write an equation where w is the number of weeks Caleb saves his money and t is the total amount in dollars in the cookie jar. Determine which variable is the independent variable and which is the dependent variable. Then, graph the total amount in the cookie jar for w being less than 6 weeks.

$t = 10w + 25$

The total amount, t, is the dependent variable.

The number of weeks, w, is the independent variable.

# of Weeks	Total Amount in Cookie Jar ($)
0	25
1	35
2	45
3	55
4	65
5	75

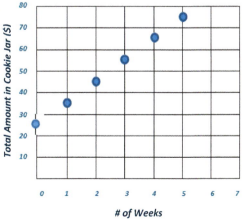

2. Kevin is taking a taxi from the airport to his home. There is a $6 flat fee for riding in the taxi. In addition, Kevin must also pay $1 per mile. Write an equation where m is the number of miles and t is the total cost in dollars of the taxi ride. Determine which variable is independent and which is dependent. Then, graph the total cost for m being less than 6 miles.

$t = 1m + 6$

The total cost in dollars, t, is the dependent variable.

The number of miles, m, is the independent variable.

# of Miles	Total Cost ($)
0	6
1	7
2	8
3	9
4	10
5	11

Total Cost of a Taxi Ride

3. Anna started with $10. She saved an additional $5 each week. Write an equation that can be used to determine the total amount saved in dollars, t, after a given number of weeks, w. Determine which variable is independent and which is dependent. Then, graph the total amount saved for the first 8 weeks.

$t = 5w + 10$

The total amount saved in dollars, t, is the dependent variable.

The number of weeks, w, is the independent variable.

# of Weeks	Total Amount ($)
0	10
1	15
2	20
3	25
4	30
5	35
6	40
7	45
8	50

Total Amount Saved

EUREKA
MATH

4. Aliyah is purchasing produce at the farmers' market. She plans to buy $10 worth of potatoes and some apples. The apples cost $1.50 per pound. Write an equation to show the total cost of the produce, where T is the total cost, in dollars, and a is the number of pounds of apples. Determine which variable is dependent and which is independent. Then, graph the equation on the coordinate plane.

$T = 1.50a + 10$

The total cost in dollars is the dependent variable. The number of pounds of apples is the independent variable.

# of Pounds of Apples	Total Cost ($)
0	10
1	11.50
2	13
3	14.50
4	16
5	17.50

Total Cost at the Farmers' Market

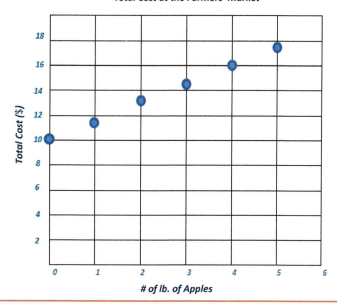

EUREKA
MATH™

Lesson 33: From Equations to Inequalities

Student Outcomes

- Students understand that an inequality with numerical expressions is either true or false. It is true if the numbers calculated on each side of the inequality sign result in a correct statement and is false otherwise.
- Students understand solving an inequality is answering the question of which values from a specified set, if any, make the inequality true.

Classwork

Example 1 (8 minutes)

Students review their work from Lesson 23 and use this throughout the lesson.

> **Example 1**
>
> What value(s) does the variable have to represent for the equation or inequality to result in a true number sentence?
> What value(s) does the variable have to represent for the equation or inequality to result in a false number sentence?
>
> a. $y + 6 = 16$
>
>   *The number sentence is true when y is 10. The sentence is false when y is any number other than 10.*
>
> b. $y + 6 > 16$
>
>   *The number sentence is true when y is any number greater than 10. The sentence is false when y is 10 or any number less than 10.*
>
> c. $y + 6 \geq 16$
>
>   *The number sentence is true when y is 10 or any number greater than 10. The sentence is false when y is a number less than 10.*
>
> d. $3g = 15$
>
>   *The number sentence is true when g is 5. The number sentence is false when g is any number other than 5.*
>
> e. $3g < 15$
>
>   *The number sentence is true when g is any number less than 5. The number sentence is false when g is 5 or any number greater than 5.*
>
> f. $3g \leq 15$
>
>   *The number sentence is true when g is 5 or any number less than 5. The number sentence is false when g is any number greater than 5.*

EUREKA
MATH™

Example 2 (12 minutes)

Students move from naming the values that make the sentence true or false to using a set of numbers and determining whether or not the numbers in the set make the equation or inequality true or false.

Example 2

Which of the following number(s), if any, make the equation or inequality true: $\{0, 3, 5, 8, 10, 14\}$?

   a. $m + 4 = 12$

        $m = 8 \; or \; \{8\}$

   b. $m + 4 < 12$

        $\{0, 3, 5\}$

- How does the answer to part (a) compare to the answer to part (b)?
  - *In part (a), 8 is the only number that will result in a true number sentence. But in part (b), any number in the set that is less than 8 will make the number sentence true.*

---

   c. $f - 4 = 2$

        None of the numbers in the set will result in a true number sentence.

   d. $f - 4 > 2$

        $\{8, 10, 14\}$

MP.6

- Is there a number that we could include in the set so that part (c) will have a solution?
  - *Yes. The number 6 will make the equation in part (c) true.*
- Would 6 be part of the solution set in part (d)?
  - *No. The 6 would not make part (d) a true number sentence because $6 - 4$ is not greater than 2.*
- How could we change part (d) so that 6 would be part of the solution?
  - *Answers will vary; If the $>$ was changed to a $\geq$, we could include 6 in the solution set.*

---

   e. $\dfrac{1}{2}h = 8$

        None of the numbers in the set will result in a true number sentence.

   f. $\dfrac{1}{2}h \geq 8$

        None of the numbers in the set will result in a true number sentence.

- Which whole numbers, if any, make the inequality in part (f) true?
  - *Answers will vary; 16 and any number greater than 16 will make the number sentence true.*

©2015 Great Minds. eureka-math.org
G6-M4-TE-B4-1.3.1-01.2016

Exercises (16 minutes)

Students practice either individually or in pairs.

Exercises

Choose the number(s), if any, that make the equation or inequality true from the following set of numbers:
$\{0, 1, 5, 8, 11, 17\}$.

1. $m + 5 = 6$

    $m = 1$ or $\{1\}$

2. $m + 5 \leq 6$

    $\{0, 1\}$

3. $5h = 40$

    $h = 8$ or $\{8\}$

4. $5h > 40$

    $\{11, 17\}$

5. $\frac{1}{2}y = 5$

    There is no solution in the set.

6. $\frac{1}{2}y \leq 5$

    $\{0, 1, 5, 8\}$

7. $k - 3 = 20$

    There is no solution in the set.

8. $k - 3 > 20$

    There is no solution in the set.

EUREKA
MATH™

Closing (3 minutes)

- In some of the equations and inequalities we worked within this lesson, none of the numbers in the given set were solutions. What does this mean? Are there numbers that will make the number sentences true that are not in the set?

    □ *None of the numbers in the set resulted in a true number sentence. However, there are numbers that could make the number sentence true. For example, in Exercise 5, $y = 10$ would make a true number sentence but was not included in the given set of numbers.*

- Is it possible for every number in a set to result in a true number sentence?

    □ *Yes, it is possible. For example, if the inequality says $x > 5$ and all the numbers in the set are greater than 5, then all the numbers in the set will result in a true number sentence.*

- Consider the equation $y + 3 = 11$ and the inequality $y + 3 < 11$. How does the solution to the equation help you determine the solution set to the inequality?

    □ *In the equation $y + 3 = 11$, $y = 8$ will result in a true number sentence. In the inequality, we want $y + 3$ to be a value less than 11. So, the numbers that will make it true must be less than 8.*

Exit Ticket (6 minutes)

©2015 Great Minds. eureka-math.org
G6-M4-TE-B4-1.3.1-01.2016

Name _____   Date _____

Lesson 33: From Equations to Inequalities

Exit Ticket

Choose the number(s), if any, that make the equation or inequality true from the following set of numbers:
$\{3, 4, 7, 9, 12, 18, 32\}$.

1. $\frac{1}{3}f = 4$

2. $\frac{1}{3}f < 4$

3. $m + 7 = 20$

4. $m + 7 \geq 20$

©2015 Great Minds. eureka-math.org
G6-M4-TE-B4-1.3.1-01.2016

EUREKA
MATH™

Exit Ticket Sample Solutions

Choose the number(s), if any, that make the equation or inequality true from the following set of numbers: $\{3, 4, 7, 9, 12, 18, 32\}$.

1. $\frac{1}{3}f = 4$

    $f = 12$ or $\{12\}$

2. $\frac{1}{3}f < 4$

    $\{3, 4, 7, 9\}$

3. $m + 7 = 20$

    There is no number in the set that will make this equation true.

4. $m + 7 \geq 20$

    $\{18, 32\}$

Problem Set Sample Solutions

Choose the number(s), if any, that make the equation or inequality true from the following set of numbers: $\{0, 3, 4, 5, 9, 13, 18, 24\}$.

1. $h - 8 = 5$

    $h = 13$ or $\{13\}$

2. $h - 8 < 5$

    $\{0, 3, 4, 5, 9\}$

3. $4g = 36$

    $g = 9$ or $\{9\}$

4. $4g \geq 36$

    $\{9, 13, 18, 24\}$

5. $\frac{1}{4}y = 7$

    There is no number in the set that will make this equation true.

6. $\frac{1}{4}y > 7$

    There is no number in the set that will make this inequality true.

©2015 Great Minds. eureka-math.org
G6-M4-TE-B4-1.3.1-01.2016

7. $m - 3 = 10$

$m = 13 \ or \ \{13\}$

8. $m - 3 \leq 10$

$\{0, 3, 4, 5, 9, 13\}$

EUREKA
MATH

Lesson 34: Writing and Graphing Inequalities in Real-World Problems

Student Outcomes

- Students recognize that inequalities of the form $x < c$ and $x > c$, where x is a variable and c is a fixed number, have infinitely many solutions when the values of x come from a set of rational numbers.

Classwork

Example 1 (10 minutes)

Begin with a discussion of what each of these statements means. Have students share possible amounts of money that could fit the given statement to build toward a graph and an inequality.

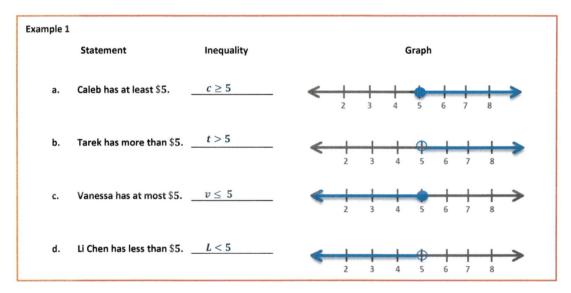

Example 1

	Statement	Inequality	Graph
a.	Caleb has at least $5.	$c \geq 5$	
b.	Tarek has more than $5.	$t > 5$	
c.	Vanessa has at most $5.	$v \leq 5$	
d.	Li Chen has less than $5.	$L < 5$	

MP.4

- How much money could Caleb have?
  - *He could have $5, $5.01, $5.90, $6, $7, $8, $9, …. More simply, he could have $5 or any amount greater than $5.*
- How would we show this as an inequality?
  - *$c \geq 5$, where c is the amount of money that Caleb has in dollars*
- What numbers on the graph do we need to show as a solution?
  - *5 is a solution and everything to the right.*
- Because we want to include 5 in the solution, we will draw a solid circle over the 5 and then an arrow to the right to show that all the numbers 5 and greater are part of the solution.

- How does the statement about Tarek differ from the statement about Caleb?

  ▫ *Tarek has more than $5, but he cannot have exactly $5, where Caleb might have had exactly $5.*

- So, how would we show this as an inequality?

  ▫ *$t > 5$, where t is the amount of money Tarek has in dollars*

- When we graph the inequality for Tarek, we still want a circle on the 5, but this time it will not be solid to show that 5 is not included in the solution.

- What does "at most" mean in Vanessa's example?

  ▫ *Vanessa could have $5 but no more than 5. So, she could have less than $5, including $4, $3, $2, $1, $0, or even a negative amount if she owes someone money.*

- How would we write this as an inequality?

  ▫ *$v \leq 5$, where v is the amount of money Vanessa has in dollars*

- How would you show this on the graph?

  ▫ *We would put a circle on the 5 and then an arrow toward the smaller numbers.*

- Would we have a solid or an open circle?

  ▫ *It would be solid to show that 5 is part of the solution.*

- Would the inequality and graph for Li Chen be the same as Vanessa's solution? Why or why not?

  ▫ *No. They would be similar but not exactly the same. Li Chen cannot have $5 exactly. So, the circle in the graph would be open, and the inequality would be $L < 5$, where L represents the amount of money Li Chen has in dollars.*

MP.4

Example 2 (5 minutes)

Example 2

Kelly works for Quick Oil Change. If customers have to wait longer than 20 minutes for the oil change, the company does not charge for the service. The fastest oil change that Kelly has ever done took 6 minutes. Show the possible customer wait times in which the company charges the customer.

$6 \leq x \leq 20$

- How is this example different from the problems in Example 1?

  ▫ *This one is giving a range of possible values. The number of minutes he takes to change the oil should be somewhere between two values instead of greater than just one or less than just one.*

- Let's start with the first bit of information. What does the second sentence of the problem tell us about the wait times for paying customers?

  ▫ *The oil change must take 20 minutes or less.*

- How would we show this on a number line?

  ▫ *Because 20 minutes is part of the acceptable time limit, we will use a solid circle and shade to the left.*

- Now, let's look at the other piece of information. The fastest Kelly has ever completed the oil change is 6 minutes. What does this mean about the amount of time it will take?

  □ *This means that it will take 6 minutes or more to complete an oil change.*

- How would we show this on a number line?

  □ *Because 6 minutes is a possible amount of time, we will use a solid circle. Then, we will shade to the right.*

- Now, we need to put both of these pieces of information together to make one model of the inequality.

- How could we show both of these on one number line?

  □ *Instead of an arrow, we would have two circles, and we would shade in between.*

- Should the circles be open or solid?

  □ *Because he has to change the oil in 20 minutes or less, the 20 is part of the solution, and the circle will be closed. The 6 minutes is also part of the solution because it is an actual time that Kelly has completed the work. The circle at 6 should also be closed.*

Example 3 (5 minutes)

MP.4

> **Example 3**
>
> Gurnaz has been mowing lawns to save money for a concert. Gurnaz will need to work for at least six hours to save enough money, but he must work fewer than 16 hours this week. Write an inequality to represent this situation, and then graph the solution.
>
>
>
> $6 \leq x < 16$

- How would we represent Gurnaz working at least six hours?

  □ *"At least" tells us that Gurnaz must work 6 hours or more.*

  □ $x \geq 6$

- What inequality would we use to show that he must work fewer than 16 hours?

  □ *"Fewer than" means that Gurnaz cannot actually work 16 hours. So, we will use $x < 16$.*

Exercises 1–5 (15 minutes)

Students work individually.

Exercises 1–5

Write an inequality to represent each situation. Then, graph the solution.

1. Blayton is at most 2 meters above sea level.

    $b \leq 2$, where b is Blayton's position in relationship to sea level in meters

2. Edith must read for a minimum of 20 minutes.

    $E \geq 20$, where E is the number of minutes Edith reads

3. Travis milks his cows each morning. He has never gotten fewer than 3 gallons of milk; however, he always gets fewer than 9 gallons of milk.

    $3 \leq x < 9$, where x represents the gallons of milk

4. Rita can make 8 cakes for a bakery each day. So far, she has orders for more than 32 cakes. Right now, Rita needs more than four days to make all 32 cakes.

    $x > 4$, where x is the number of days Rita has to bake the cakes

5. Rita must have all the orders placed right now done in 7 days or fewer. How will this change your inequality and your graph?

    $4 < x \leq 7$

    Our inequality will change because there is a range for the number of days Rita has to bake the cakes. The graph has changed because Rita is more limited in the amount of time she has to bake the cakes. Instead of the graph showing any number larger than 4, the graph now has a solid circle at 7 because Rita must be done baking the cakes in a maximum of 7 days.

Possible Extension Exercises 6–10

The following problems combine the skills used to solve equations in previous lessons within this module and inequalities.

Possible Extension Exercises 6–10

6. Kasey has been mowing lawns to save up money for a concert. He earns $15 per hour and needs at least $90 to go to the concert. How many hours should he mow?

$$15x \geq 90$$
$$\frac{15x}{15} \geq \frac{90}{15}$$
$$x \geq 6$$

Kasey will need to mow for 6 or more hours.

EUREKA MATH

7. Rachel can make 8 cakes for a bakery each day. So far, she has orders for more than 32 cakes. How many days will it take her to complete the orders?

$$8x > 32$$
$$\frac{8x}{8} > \frac{32}{8}$$
$$x > 4$$

Rachel will need to work more than 4 days.

8. Ranger saves $70 each week. He needs to save at least $2,800 to go on a trip to Europe. How many weeks will he need to save?

$$70x \geq 2800$$
$$\frac{70x}{70} \geq \frac{2800}{70}$$
$$x \geq 40$$

Ranger needs to save for at least 40 weeks.

9. Clara has less than $75. She wants to buy 3 pairs of shoes. What price shoes can Clara afford if all the shoes are the same price?

$$3x < 75$$
$$\frac{3x}{3} < \frac{75}{3}$$
$$x < 25$$

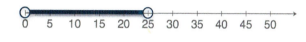

Clara can afford shoes that are greater than $0 and less than $25.

10. A gym charges $25 per month plus $4 extra to swim in the pool for an hour. If a member only has $45 to spend each month, at most how many hours can the member swim?

$$4x + 25 \leq 45$$
$$4x + 25 - 25 \leq 45 - 25$$
$$4x \leq 20$$
$$\frac{4x}{4} \leq \frac{20}{4}$$
$$x \leq 5$$

The member can swim in the pool for 5 hours. However, we also know that the total amount of time the member spends in the pool must be greater than or equal to 0 hours because the member may choose not to swim.

$$0 \leq x \leq 5$$

Closing (5 minutes)

- How are inequalities different from equations?
    - *Inequalities can have a range of possible values that make the statement true, where equations do not.*
- Does the phrase "at most" refer to being less than or greater than something? Give an example to support your answer.
    - *"At most" means that you can have that amount or less than that amount. You cannot go over. My mom says that I can watch at most 3 TV shows after I do my homework. This means that I can watch 3 or fewer than 3 TV shows.*

Exit Ticket (5 minutes)

Name _____   Date _____

Lesson 34: Writing and Graphing Inequalities in Real-World Problems

Exit Ticket

For each question, write an inequality. Then, graph your solution.

1. Keisha needs to make at least 28 costumes for the school play. Since she can make 4 costumes each week, Keisha plans to work on the costumes for at least 7 weeks.

2. If Keisha has to have the costumes complete in 10 weeks or fewer, how will our solution change?

EUREKA MATH

Exit Ticket Sample Solutions

For each question, write an inequality. Then, graph your solution.

1. Keisha needs to make at least 28 costumes for the school play. Since she can make 4 costumes each week, Keisha plans to work on the costumes for at least 7 weeks.

    $x \geq 7$

    Keisha should plan to work on the costumes for 7 or more weeks.

    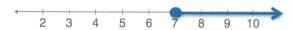

2. If Keisha has to have the costumes complete in 10 weeks or fewer, how will our solution change?

    Keisha had 7 or more weeks in Problem 1. It will still take her at least 7 weeks, but she cannot have more than 10 weeks.

    $7 \leq x \leq 10$

Problem Set Sample Solutions

Write and graph an inequality for each problem.

1. At least 13

    $x \geq 13$

2. Less than 7

    $x < 7$

3. Chad will need at least 24 minutes to complete the 5K race. However, he wants to finish in under 30 minutes.

    $24 \leq x < 30$

4. Eva saves $60 each week. Since she needs to save at least $2,400 to go on a trip to Europe, she will need to save for at least 40 weeks.

    $x \geq 40$

EUREKA
MATH™

5. Clara has $100. She wants to buy 4 pairs of the same pants. Due to tax, Clara can afford pants that are less than $25.

    Clara must spend less than $25, but we also know that Clara will spend more than $0 when she buys pants at the store.

    $0 < x < 25$

6. A gym charges $30 per month plus $4 extra to swim in the pool for an hour. Because a member has just $50 to spend at the gym each month, the member can swim at most 5 hours.

    The member can swim in the pool for 5 hours. However, we also know that the total amount of time the member spends in the pool must be greater than or equal to 0 hours because the member may choose not to swim.

    $0 \leq x \leq 5$

EUREKA
MATH™

Name _____   Date _____

1. Gertrude is deciding which cell phone plan is the best deal for her to buy. Super Cell charges a monthly fee of \$10 and also charges \$0.15 per call. She makes a note that the equation is $M = 0.15C + 10$, where M is the monthly charge, in dollars, and C is the number of calls placed. Global Cellular has a plan with no monthly fee but charges \$0.25 per call. She makes a note that the equation is $M = 0.25C$, where M is the monthly charge, in dollars, and C is the number of calls placed. Both companies offer unlimited text messages.

   a. Make a table for both companies showing the cost of service, M, for making from 0 to 200 calls per month. Use multiples of 20.

| | Cost of Services, M, in Dollars | |
Number of Calls, C	Super Cell $M = 0.15C + 10$	Global Cellular $M = 0.25C$

b. Construct a graph for the two equations on the same graph. Use the number of calls, C, as the independent variable and the monthly charge, in dollars, M, as the dependent variable.

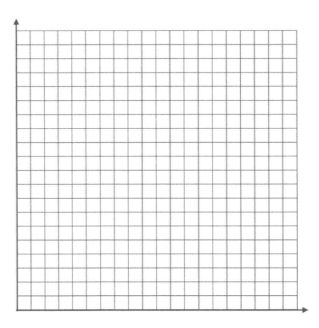

c. Which cell phone plan is the best deal for Gertrude? Defend your answer with specific examples.

EUREKA
MATH™

2. Sadie is saving her money to buy a new pony, which costs $600. She has already saved $75. She earns $50 per week working at the stables and wonders how many weeks it will take to earn enough for a pony of her own.

   a. Make a table showing the week number, W, and total savings, in dollars, S, in Sadie's savings account.

Number of Weeks											
Total Savings (in dollars)											

   b. Show the relationship between the number of weeks and Sadie's savings using an expression.

   c. How many weeks will Sadie have to work to earn enough to buy the pony?

Module 4:   Expressions and Equations

379

3. The elevator at the local mall has a weight limit of 1,800 pounds and requires that the maximum person allowance be no more than nine people.

    a. Let x represent the number of people. Write an inequality to describe the maximum allowance of people allowed in the elevator at one time.

    b. Draw a number line diagram to represent all possible solutions to part (a).

    c. Let w represent the amount of weight, in pounds. Write an inequality to describe the maximum weight allowance in the elevator at one time.

    d. Draw a number line diagram to represent all possible solutions to part (c).

4. Devin's football team carpools for practice every week. This week is his parents' turn to pick up team members and take them to the football field. While still staying on the roads, Devin's parents always take the shortest route in order to save gasoline. Below is a map of their travels. Each gridline represents a street and the same distance.

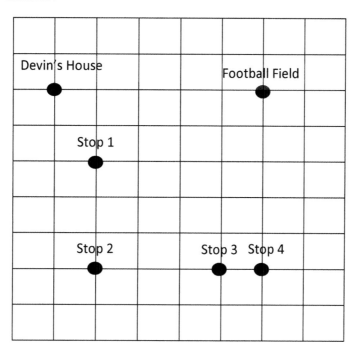

Devin's father checks his mileage and notices that he drove 18 miles between his house and Stop 3.

a. Create an equation, and determine the amount of miles each gridline represents.

b. Using this information, determine how many total miles Devin's father will travel from home to the football field, assuming he made every stop. Explain how you determined the answer.

c. At the end of practice, Devin's father dropped off team members at each stop and went back home. How many miles did Devin's father travel all together?

5. For a science experiment, Kenneth reflects a beam off a mirror. He is measuring the missing angle created when the light reflects off the mirror. (Note: The figure is not drawn to scale.)

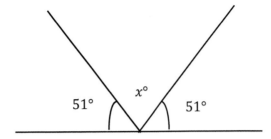

Use an equation to determine the missing angle, labeled x in the diagram.

EUREKA
MATH

A Progression Toward Mastery

Assessment Task Item		STEP 1 Missing or incorrect answer and little evidence of reasoning or application of mathematics to solve the problem.	STEP 2 Missing or incorrect answer but evidence of some reasoning or application of mathematics to solve the problem.	STEP 3 A correct answer with some evidence of reasoning or application of mathematics to solve the problem, OR an incorrect answer with substantial evidence of solid reasoning or application of mathematics to solve the problem.	STEP 4 A correct answer supported by substantial evidence of solid reasoning or application of mathematics to solve the problem.
1	a 6.EE.C.9	Student is not able to make a table with multiples of 20 calls or cannot calculate the monthly charge based on the number of calls.	Student correctly completes the Number of Calls column but is not able to use the equations to accurately complete the other two columns.	Student accurately calculates one column, but the other two columns have errors.	Student accurately calculates all of the columns.
	b 6.EE.C.9 6.EE.B.6	Student is not able to graph the data from the table.	Student attempts to graph the data from the table. Several mistakes or omissions are present.	Student graphs the data from the table but has minor mistakes or omissions. Student includes at least three of the four criteria of a perfect graph.	Student graphs the data from the table without error (i.e., graph is titled, axes are labeled, units are included on axes, and points are plotted accurately).
	c 6.EE.C.9	Student cannot conclude which plan is better or chooses a plan without evidentiary support.	Student chooses one plan or the other, showing support for that plan. Student does not recognize that the best deal depends on the number of calls placed in a month.	Student concludes that the best deal depends on the number of calls placed in a month. Student does not describe a complete analysis.	Student concludes that the best deal depends on the number of calls placed in a month. The break-even point, 100 calls costing $25, is identified. Answer specifically states that Super Cell is the better deal if the number of calls per month is < 100, and Global Cellular is the better deal if the number of calls per month is > 100.

2	a 6.EE.C.9	Student is not able to make the table or attempts to make the table but has many errors.	Student is able to make the table and calculate most rows accurately. Compounding errors may be present. Titles or variables may be missing.	Student accurately makes the table and accurately computes the total savings for each week. Titles or variables may be missing. OR Student begins with 50 and 75 every week.	Student accurately makes the table and accurately computes the total savings for each week. Titles and variables are present.
	b 6.EE.C.9	Student is not able to write the expression.	Student attempts to write the expression but is inaccurate (perhaps writing $75W + 50$).	Student writes the expression $50W + 75$ but does not include a description of what the variable represents.	Student accurately writes the expression $50W + 75$ and includes a description of what the variable represents.
	c 6.EE.C.9	Student cannot make a conclusion about how many weeks are needed for the purchase, or an answer is provided that is not supported by the student's table.	Student concludes that some number of weeks other than 11 weeks will be needed for the purchase.	Student concludes that 11 weeks of work will be needed for the purchase.	Student concludes that 11 weeks of work will be needed for the purchase and that Sadie will have $625, which is $25 more than the cost of the pony.
3	a 6.EE.B.8	Student does not write an inequality or tries writing the inequality with incorrect information (e.g., 1,800).	Student writes the inequality $x \geq 9$ or $x > 9$.	Student writes the inequality $x < 9$ because student does not realize there can be 9 people on the elevator.	Student writes the inequality $0 \leq x \leq 9$.
	b 6.EE.C.9	Student does not draw a number line or draws a line but does not indicate $0 \leq x \leq 9$.	Student draws an accurate number line and uses either a line segment or discrete symbols but does not include 0 and/or 9 in the solution set.	Student draws a number line but uses a line segment to indicate continuous points $0 \leq x \leq 9$.	Student draws an accurate number line, using discrete symbols indicating whole numbers from 0 to 9.
	c 6.EE.B.8	Student does not write an inequality or tries writing the inequality with incorrect information (e.g., 9).	Student writes the inequality $W \geq 1,800$ or $W > 1,800$.	Student writes the inequality $W < 1,800$ because student does not realize there can be 1,800 pounds on the elevator.	Student writes the inequality $0 \leq W \leq 1,800$.
	d 6.EE.C.9	Student does not draw a number line or draws a line but does not indicate $0 \leq W \leq 1,800$.	Student draws an accurate number line and uses either a line segment or discrete symbols but does not include 0 and/or 1,800 in the solution set.	Student draws a number line but uses discrete symbols indicating whole numbers from 0 to 1,800 are in the solution set $0 \leq W \leq 1,800$.	Student draws an accurate number line, using a line segment indicating all numbers from 0 to 1,800 are in the solution set $0 \leq W \leq 1,800$.

Module 4:   Expressions and Equations

EUREKA MATH™

4	a 6.EE.B.7 6.EE.C.9 6.EE.B.5 6.EE.B.6	Student is unable to create an equation and is unable to determine the amount of miles Devin's father traveled between their house and Stop 3.	Student is able to determine the amount of miles Devin's father traveled between their house and Stop 3 but does not write an equation.	Student creates an equation, $9G = 18$ mi, but does not use the equation to determine the amount of miles Devin's father traveled between their house and Stop 3.	Student creates an equation and uses it to determine the number of miles Devin's father traveled between their house and Stop 3. Let G represent the number of gridlines passed on the map. $$9G = 18 \text{ mi}$$ $$\frac{G}{9} = \frac{18 \text{ mi}}{9}$$ $$G = 2 \text{ mi}$$
	b 6.EE.A.2 6.EE.B.6 6.EE.B.6	Student does not describe how the answer was derived or leaves the answer blank.	Student determines the correct distance (30 miles) but does not explain how the answer was determined or offers an incomplete explanation.	Student inaccurately counts the intersections passed (15) but accurately applies the equation with the incorrect count. Explanation is correct and clear.	Student accurately counts the intersections passed (15) and applies that to the correct equation: $15 \cdot 2$ mi $= 30$ mi.
	c 6.EE.C.9 6.EE.B.5 6.EE.B.6	Student answer does not indicate a concept of round-trip distance being double that of a one-way trip.	Student does not double the correct one-way trip distance (30 miles) from part (b) or doubles the number incorrectly.	Student does not use the one-way trip distance (30 miles) from part (b) but counts the blocks for the round trip.	Student doubles the correct one-way trip distance (30 miles) from part (b) to arrive at the correct round-trip distance (60 miles).
5	6.EE.B.5 6.EE.B.6 6.EE.B.7	Student does not show any of the steps necessary to solve the problem or simply answers 51°.	Student adds $51° + 51°$ to arrive at 102° but does not subtract this from 180° to find the missing angle.	Student correctly finds the missing angle, 78°, showing clearly the steps involved but does not use an equation. OR Student makes an arithmetic error, but clear evidence of conceptual understanding is evident.	Student correctly finds the missing angle, 78°, by using an equation and clearly showing the steps involved. Student might reference the terms *supplementary angles* or *straight angles* or start with $51° + 51° + x° = 180°$ before solving it correctly.

Module 4:   Expressions and Equations

385

©2015 Great Minds. eureka-math.org
G6-M4-TE-B4-1.3.1-01.2016

Name _____   Date _____

1. Gertrude is deciding which cell phone plan is the best deal for her to buy. Super Cell charges a monthly fee of $10 and also charges $0.15 per call. She makes a note that the equation is $M = 0.15C + 10$, where M is the monthly charge, in dollars, and C is the number of calls placed. Global Cellular has a plan with no monthly fee but charges $0.25 per call. She makes a note that the equation is $M = 0.25C$, where M is the monthly charge, in dollars, and C is the number of calls placed. Both companies offer unlimited text messages.

   a. Make a table for both companies showing the cost of service, M, for making from 0 to 200 calls per month. Use multiples of 20.

Number of Calls, C	Cost of Services, M, in Dollars	
	Super Cell $M = 0.15C + 10$	Global Cellular $M = 0.25C$
0	10	0
20	13	5
40	16	10
60	19	15
80	22	20
100	25	25
120	28	30
140	31	35
160	34	40
180	37	45
200	40	50

©2015 Great Minds. eureka-math.org
G6-M4-TE-B4-1.3.1-01.2016

EUREKA MATH

b. Construct a graph for the two equations on the same graph. Use the number of calls, C, as the independent variable and the monthly charge, in dollars, M, as the dependent variable.

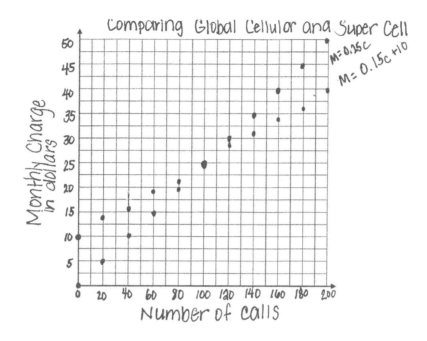

c. Which cell phone plan is the best deal for Gertrude? Defend your answer with specific examples.

The best deal depends on the number of calls placed in a month. The break even point is 100 calls per month. Super Cell is a better deal if the number of monthly calls is less than 100. Global Cellular is a better deal if the number of monthly calls is greater than 100.

2. Sadie is saving her money to buy a new pony, which costs $600. She has already saved $75. She earns $50 per week working at the stables and wonders how many weeks it will take to earn enough for a pony of her own.

   a. Make a table showing the week number, W, and total savings, in dollars, S, in Sadie's savings account.

   b. Show the relationship between the number of weeks and Sadie's savings using an expression.

Number of Weeks	1	2	3	4	5	6	7	8	9	10	11	12
Total Savings (in dollars)	125	175	225	275	325	375	425	475	525	575	625	675

$$50w + 75$$

   c. How many weeks will Sadie have to work to earn enough to buy the pony?

   If Sadie works 11 weeks, she will earn $625, which is $25 more than the cost of the pony.

EUREKA MATH

3. The elevator at the local mall has a weight limit of 1,800 pounds and requires that the maximum person allowance be no more than nine people.

   a. Let x represent the number of people. Write an inequality to describe the maximum allowance of people allowed in the elevator at one time.

   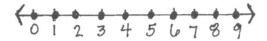

   $$0 \leq x \leq 9$$

   b. Draw a number line diagram to represent all possible solutions to part (a).

   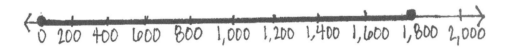

   c. Let w represent the amount of weight, in pounds. Write an inequality to describe the maximum weight allowance in the elevator at one time.

   $$0 \leq w \leq 1,800$$

   d. Draw a number line diagram to represent all possible solutions to part (c).

4. Devin's football team carpools for practice every week. This week is his parents' turn to pick up team members and take them to the football field. While still staying on the roads, Devin's parents always take the shortest route in order to save gasoline. Below is a map of their travels. Each gridline represents a street and the same distance.

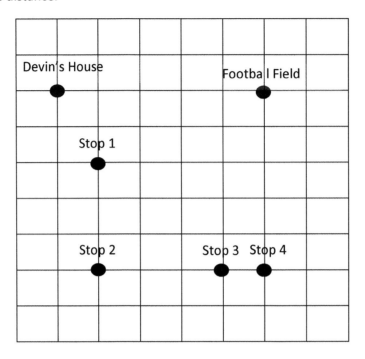

Devin's father checks his mileage and notices that he drove 18 miles between his house and Stop 3.

a. Create an equation, and determine the amount of miles each gridline represents.

Let G represent the number of gridlines passed on the map. $9G=18$ $\dfrac{9G}{9}=\dfrac{18}{9}$ $G=2\text{ miles}$.

b. Using this information, determine how many total miles Devin's father will travel from home to the football field, assuming he made every stop. Explain how you determined the answer.

$15\,G=\text{miles}$
$15\,(2\text{ miles})=30\text{ miles}$

c. At the end of practice, Devin's father dropped off team members at each stop and went back home. How many miles did Devin's father travel all together?

$30\,G=\text{miles}$
$30\,(2\text{ miles})=60\text{ miles}$

EUREKA
MATH

5. For a science experiment, Kenneth reflects a beam off a mirror. He is measuring the missing angle created when the light reflects off the mirror. (Note: The figure is not drawn to scale.)

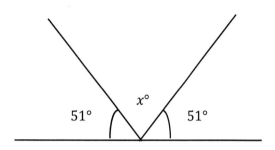

Use an equation to determine the missing angle, labeled x in the diagram.

A straight angle measures $180°$.

$$51° + x° + 51° = 180°$$
$$x° + 102° = 180°$$
$$x° + 102° - 102° = 180° - 102°$$
$$x° = 78°$$

This page intentionally left blank